41

OXFORD READINGS IN PHILOSOPHY

Series Editor G. J. Warnock

THE PHILOSOPHY OF MATHEMATICS

Also published in this series

Theories of Ethics edited by Philippa Foot
Knowledge and Belief edited by A. Phillips Griffiths
The Philosophy of Science edited by P. H. Nidditch
The Theory of Meaning edited by G. H. R. Parkinson
Political Philosophy edited by Anthony Quinton
Philosophical Logic edited by P. F. Strawson
The Philosophy of Perception edited by G. J. Warnock
The Philosophy of Action edited by Alan R. White

Other volumes are in preparation

THE PHILOSOPHY OF MATHEMATICS

Edited by
JAAKKO HINTIKKA

OXFORD UNIVERSITY PRESS
1969

*Oxford University Press, Ely House, London W.*1
GLASGOW NEW YORK TORONTO MELBOURNE WELLINGTON
CAPE TOWN SALISBURY IBADAN NAIROBI LUSAKA ADDIS ABABA
BOMBAY CALCUTTA MADRAS KARACHI LAHORE DACCA
KUALA LUMPUR SINGAPORE HONG KONG TOKYO

SET BY SPOTTISWOODE, BALLANTYNE & CO. LTD.
LONDON AND COLCHESTER
PRINTED IN GREAT BRITAIN BY
FLETCHER AND SON LTD, NORWICH

CONTENTS

INTRODUCTION

: may seem strange to reprint the papers collected here under the title 'The hilosophy of Mathematics'. In the majority of them, no specifically philo-›phical problems seem to be discussed.

Although this impression is not unfounded, in the present circumstances .e papers appearing here nevertheless exemplify the kind of reading which judge to be by far the most useful for a student of the philosophy of athematics. In these days, the worthwhile articles and books devoted to e philosophy of mathematics in the narrow sense of the word 'philosophy' e few, and their quality frequently leaves a great deal to be desired. Often e reader soon discovers that such a book or article is not based on any-ing like the adequate acquaintance with the large and important body of *aterials* which is indispensable for any future philosophy of mathematics, mely the literature of symbolic logic and foundational studies. It is not cely that any substantial progress can be made in the genuinely philo-phical study of mathematics without using the concepts and results veloped in this literature to a much greater extent than has happened so r. However true it may be that no philosophical problems are solved by sults in logic, not to use the wealth of material which is available here nply makes the discussion as unrealistic as a discussion of legal philo-phy would be if it were not based on any knowledge of the actual oper-ion of the law. Thus the most useful contribution that a book of readings n make in this area is to ensure that some of the recent results in logic d foundational studies are accessible to an interested reader. To do this the first aim of the present volume.

What, then, are these philosophically relevant insights that a philos-her of mathematics ought to be interested in? This question cannot be swered adequately within the confines of this introduction, for an answer ould amount to a survey of an extensive, rapidly growing, and complex ld. Furthermore, about as good a survey as anyone can hope to find ready exists in Andrzej Mostowski's volume *Thirty Years of Foundational udies*, listed in the bibliography at the end of this volume. (All other ferences in this introduction are also to works listed in the same biblio-aphy, unless otherwise specified.)

What can be done here is to try to supply enough background for the ections reprinted in this volume to enable a reader to make use of them. e following informal sketch cannot help being loose and inaccurate. I ve tried mainly to indicate some of the most interesting interconnections between the different types of recent work represented here, and also t mention a few recent developments which are not represented in o selections.

Probably the most important large-scale recent development in logic an foundational studies is the growth of the theory which used to be known a logical semantics but which is now usually referred to as *model theory.* can be characterized, very roughly, as a study of the interrelations of language—usually a precise logical symbolism—and the 'reality' th language represents. The basic questions concern the conditions in which formula can be said to be true in a structure or 'model'. Simple as th question may seem, it soon leads to interesting problems. One of the concerns the class of sentences true in *all* models—the class of logic truths, as they are usually called. Can this class be represented as the th orems of some axiomatic system? This completeness question is answer affirmatively for first-order languages (quantificational languages). T first proof was given by Gödel in the early thirties. (See Kurt Göd 'Die Vollständigkeit der Axiome des logischen Funktionenkalküls', *Mona hefte für Mathematik und Physik*, Vol. 37 (1930), pp. 349–60.) One of t most important subsequent proofs is given in Henkin's first paper reprint in the present volume. Henkin's proof is interesting for several reasons, o of which is his strategy of forming a sequence of certain maximal consiste sets of sentences whose union can then be seen to be usable as a model which all its members are true. Certain sets of formulas thus exemplify t very structure which these formulae impose on the models in which th are true.

It is known that no similar proof exists in higher-order logics. Logi truths about arbitrary sets, unlike logical truths about arbitrary individua cannot be axiomatized. This follows from Gödel's incompleteness proof f arithmetic, for all the requisite arithmetical assumptions are easily e pressed in terms of higher-order logics. Most of the purportedly popu discussions of Gödel's famous result are somewhat unsatisfactory, ho ever. One of the best self-contained expositions seems to be Raymo Smullyan's paper reprinted here. The whole complex of problems conce ing Gödel's proof of the impossibility of a consistency proof of a syst containing arithmetic within the same system is discussed in a wider p spective in Feferman's salient paper 'Arithmetization of Metamat matics'.

However, Henkin's other paper reprinted here shows that completen can be reached even in higher-order logics if we are willing to admit 'n standard models' in which quantification over 'all sets' no longer refers all arbitrary subsets of our domain of individuals but only to all memb f some 'arbitrary' subset of the power set of the set of individuals; and milarly for pairs, triples, etc. These subsets cannot be completely arbitrary, owever, for they must be rich enough to allow the interpretation of all our ›rmulas (e.g. second-order formulas). This is a rather strong requirement, ›r it means *inter alia.*closure with respect to the 'projective' operations that ›rrespond to different kinds of quantification and not just with respect to ıe Boolean operations. If this fact is overlooked, Henkin's completeness roof for type theory may give the impression that higher-order logics are uch simpler than they really are. Their actual difficulty is illustrated, for ‹ample, by the complexity of the problem of extending the Gentzen-Ierbrand techniques (to be mentioned later) from first-order logic to igher-order logics. Takeuti's problem which is discussed in Schütte, yntactical and Semantical Properties', and which has only recently been artially solved (by Tait, Prawitz, and others) is a case in point.

An intriguing aspect of the completeness and incompleteness results is ıat one of their starting-points (viz. our concept of what constitutes com-leteness) is inevitably an idea which can perhaps be formulated in naïve ‹t-theoretical terms but which either is not formulated axiomatically to ‹gin with or which (in the case of incompleteness) cannot even possibly be › formulated. Yet concepts of this kind are most interesting. We seem to ave many clear intuitions concerning them, and it is important to develop ays of handling them. Some aspects of this situation are discussed in our first lection from Kreisel, entitled 'Informal Rigour and Completeness Proofs'.

One might describe the phenomenon of incompleteness by saying that ıe cannot axiomatically rule out all non-standard models and catch only ıe intended 'standard' ones. The inevitable presence of non-standard odels can occasionally be turned into a blessing, however, as demon-rated by Abraham Robinson's non-standard models for analysis. These odels serve to vindicate some of the locutions and ideas of the old 'meta-ıysics of the calculus' which meanwhile have been relegated to the status hopelessly loose heuristic ideas. Even infinitesimals, the bugbear of every troductory calculus course, can thus be made perfectly respectable. Here one of several recent developments that demonstrate strikingly the levance of current work to the traditional issues in the philosophy of athematics.

Completeness and incompleteness results are only samples, however, of e variety of results that have been obtained in model theory concerning hat can or cannot be expressed in different kinds of language and con-rning the variety of models that a theory can or cannot have. These sights into the limits of what can be expressed in different kinds of nguages are among the most important recent developments in the whole area of logic and foundations, and deserve more attention than has bee given to them by philosophers. The only result that has provoked muc philosophical discussion is the Löwenheim–Skolem result that eve axiomatizable first-order theory has a countable model. (For its varian and generalizations, see Vaught's 1964 paper.) The fact that this includ systems of axiomatic set theory in which the existence of uncountable se can be proved is but a further instance of the difficulty of ruling out u desirable 'non-standard' models axiomatically. Other results are not muc less interesting. A good example of them is Vaught's survey of models complete theories.

As illustrated by the Löwenheim–Skolem theorem and by Gödel's com pleteness and incompleteness results, many of the earlier basic results modern logic belong essentially to model theory. However, much of th systematic development of this area is relatively recent, often deriving i inspiration from Tarski ('Contributions to the Theory of Models', etc.) ar Abraham Robinson.

The development of model theory has not remained without repercussio for the deductive techniques of first-order logic. The basic insights go ba to Herbrand and Gentzen. Their techniques were gradually adopted in tl form of 'natural deduction methods' in many standard expositions, but interest in the underlying semantical ideas was not revived on a larg scale until 1955. Around that time, Hintikka, Beth, and Schütte (int alia) published treatments of first-order logic in which the proof of logical truth of a first-order formula can be thought of as an unsuccessf attempt to describe a counter-example to it. These counter-example d scriptions, reminiscent of Henkin's completeness proof, could even thought of as actual *models* in which the formula in question would be fal The essential simplification lies in the use of simpler descriptions of t 'possible worlds' which would serve as counter-models—descriptio simpler than Henkin's very extensive maximal consistent sets. (Tl simplification is present also in Hintikka's method, contrary to the i pression which the concluding remarks of Beth's paper in the prese volume might give.) It is clear, however, that similar ideas were alrea inspiring Herbrand, although they were not developed systematically at t time.

Beth's paper included here is the first, and perhaps the freshest, descri tion of the approach which he calls the method of semantical tableaux. completeness proof is sketched for this method in his paper, and a numb of suggestive comparisons with other issues in logic and the philosophy mathematics are presented.

The basic idea of thinking of proofs and disproofs as attempted mo nstructions seems to me of great philosophical interest and importance. e Herbrand-type point of view has in any case led to fruitful and interest- g consequences especially in the direction of Craig's interpolation lemma roved in his 1957 papers).

From the same point of view, a treatment of intuitionistic logic was given 1956 by Beth. It was commented on by Kreisel and Dyson and carried rther by Kripke. Kripke's starting-point in his 1964 paper was his earlier odel-theoretical treatment of modal logics along the same lines. Like so any ideas underlying the work of the last few decades, the first ideas in s direction go back to Gödel's early papers. In the paper entitled 'Eine terpretation des intuitionistischen Aussagenkalküls' and translated into glish here, the connection between intuitionistic and modal logics is inted out for the first time. Some of the other suggestions made in Gödel's per, notably his remarks on what happens if 'B' is interpreted as express- g provability in some particular logical system, are developed further by chard Montague in his discussion of 'Syntactical Treatments of odality', *Acta Philosophica Fennica*, Vol. 16 (1963), pp. 153–67.

The model-theoretical discussions by Beth and Kripke of intuitionistic gic do not decide the interpretation of intuitionistic mathematics, how- er: this remains an interesting and lively area for further work and even essentially new ideas. An interested reader will find a rich lode of sug- stions in those writings (or parts or writings) of Georg Kreisel which are t printed here.

The idea that disproofs can be thought of as attempted model construc- ns calls one's attention to the broader problem of the relation of formulas d their models. These relations turn largely on the concepts of truth and isfaction. These concepts are normally explained in fairly strong set- eoretical terms. Can these concepts be varied in suitable ways? Can they explained so as to rule out already at this stage some of the non-standard odels? No satisfactory answers to these questions are likely to be found the literature. Can formulae (sentences) and models be related to each ner in some ways which are closer to our actual methods of relating them practice' than are the set-theoretical methods usually employed? In- esting suggestions, couched in game-theoretical terms, have been made sentially in this direction by Lorenzen. (In addition to the paper listed in r bibliography, see W. Stegmüller, 'Remarks on the Completeness of gical Systems Relative to the Validity-Concepts of P. Lorenzen and Lorenz', *Notre Dame Journal of Formal Logic*, Vol. 5 (1964), pp. 81–112.) closely similar suggestion is put forward by Hintikka, who also argues the importance of these ideas in understanding the actual use of quanti- ational concepts in the applications of language. (See 'Language-Games for Quantifiers', *American Philosophical Quarterly*, Supplementary Mon-graph no. 2, 1968.) These suggestions seem to have at least the merit th they make it extremely natural to consider certain types of deviations fro classical logic. Lorenzen stresses the naturalness of intuitionistic logic fro this point of view. Other deviations can be obtained by requiring suitab kinds of recursivity (computability) of the strategies to be used in the games. (What could it *mean* actually to use a non-recursive strategy in game?) There does not seem to be any unique way of doing this, howeve and we soon run into the complex of problems connected with such mo or less constructivistic ideas as Gödel's extension of the finitistic point view, the 'no counter-example interpretation' of logic and arithmetic, etc

The work represented in the rest of our selections requires fewer com ments. The first part of Feferman's paper is a lucid survey of the develo ment of one important older idea in the foundations of mathematics, v the idea of predicativity. The second part presents important results co cerning this concept. Predicative reasoning turns out to be characterize roughly, by transfinite induction up to a definite ordinal of the seco number class. (This is an interesting example of a general phenomeno frequently the strength of one's assumptions turns out to be measurable the ordinals one needs in one's work.) The details of this part of Feferma paper may be rough going for most readers of this volume, but the ma results and their importance ought to be possible to grasp without all t details. Feferman's new paper 'Autonomous Transfinite Progressions' inforces and partly replaces the older one. (It contains an improved form lation of the technical results and extends them further; it also prese a sharper appraisal of what has been accomplished and what remains to done in their area.) The reader is referred to this new paper for a discussi of recent developments in the study of predicativity.

The philosophical relevance of the theory of recursive functions ought be obvious, but nevertheless little use has been made of its potentialities philosophers. The fact that the only major conceptual idealization involv in the theory of Turing machines, as compared with real digital compute is infinite memory also illustrates the importance of this theory. Little mo can be said here, however, than to refer the reader not only to Hart Rogers' brilliant expository paper, printed here, but to his long-await book. The theory of automata without this idealizing assumption of infin memory is discussed in the already classical paper by Rabin and Scott.

Although we seem to have in the notion of a Turing machine a go explication of the idea of effective computability, the situation is not simple as it may appear. From Gödel's note on 'Eine bisher noch nic benützte Erweiterung des finiten Standpunktes' one can see that there a 'ays of extending the finitistic point of view by resorting to constructive ıtities of a higher type which are not initially given any 'mechanical' ɔunterpart. From this, a distinction between what is finitistic and what is ɔnstructive seems to ensue. Further distinctions are due to the fact that a roof of the termination of a combinatorial procedure may not itself be ɔmbinatorial—a standard objection by the intuitionists to the use of :cursive function theory to explicate their intentions.

Even more obvious than the relevance of recursive function theory is the ɔnnection between recent work and traditional problems in the case of the undations of geometry. The lines indicated in Tarski's paper have been nce followed by Schwabhäuser and by a number of other logicians and ıathematicians.

So far, I have said nothing about what is often thought of as the central ea of foundational studies, viz. set theory. Nor is any selection on set ıeory included in the present volume. There exists, however, an excellent ırvey of the most important recent results in Paul Cohen's little volume.

Little that is definitive can be said of the subject matter of axiomatic set eory in the present situation, it seems to me. It has been found that some the most important open questions (especially the status of the Con- ıuum Hypothesis) cannot be solved on the basis of the customary axioms set theory. I hope that it is fair to say that most of the different ways of :tending these axioms are either unconvincing or too weak to be of much terest. Although many interesting suggestions are being discussed, it is .rd to tell where they will lead.

The only point that I want to make here concerns the direction in which rther axioms have been sought. (This, it seems to me, might turn out to the crucial philosophical problem in this area.) These new axioms have ten been formulated as postulations of suitable 'very large' cardinal ımbers. There is a certain inner logic at work here, but one cannot help ɔndering whether some completely different way of looking at them might philosophically defensible. After all, these assumptions have in general finite purely arithmetical consequences, although we do not seem to have 'ong enough intuitions about these arithmetical consequences to say ıch about them. But are some entirely different intuitions or heuristic ɜas perhaps possible here? To illustrate the implications of this question, may be mentioned that there exists a small and possibly fruitless attempt strike out in a completely different direction, to which the wider per- ectives just mentioned nevertheless lend some interest. Mycielski has perimented with completely unorthodox assumptions formulated in me-theoretical terms which have certain very pleasant mathematical nsequences. (See J. Mycielski, 'On the Axiom of Determinateness',

Fundamenta Mathematicae, Vol. 53 (1964), and J. Mycielski and H Steinhaus, 'A Mathematical Axiom Contradicting the Axiom of Choice *Bulletin de l'Academie Polonaise des Sciences*, Ser. III, Vol. 10 (1962). Whether they will remain mere curiosities, nobody seems to know, and th nicely ordered infinite numbers have so far been too seductive for th majority of logicians to trade them for Mycielski's axiom, with its muc more disorderly consequences for the interrelations of infinite numbers.

Some philosophical suggestions of the recent work in logic and i foundational studies are made in Kreisel's stimulating paper, 'Mathe matical Logic: What Has It Done for the Philosophy of Mathematics?', o which an excerpt is reprinted here. (Some new information concerning th topics discussed there is found in the appendix to this paper, as well as i Kreisel's subsequent papers.) Some of the issues touched upon earlier in th present introduction are also discussed by Kreisel in this long essay. Othe philosophical suggestions have been made elsewhere. For instance, in 'Ar Logical Truths Analytic?' and in other papers Hintikka argues that certai distinctions one can make in first-order logic constitute the best moder explication of Kant's distinction between analytic and synthetic pro positions, at least in so far as this distinction applies to logic and mathe matics. Some of the historical basis of this claim also appears from Beth brief discussion (in Section II of his paper in the present volume) of th Aristotelian concept of *ecthesis* and its later history.

I

SEMANTIC ENTAILMENT AND FORMAL DERIVABILITY

E. W. BETH

> The aggregate of all applications of logic will not compare with the treasure of the pure theory itself. For when one has surveyed the whole subject, one will see that the theory of logic, insofar as we attain to it, is the vision and the attainment of that Reasonableness for the sake of which the Heavens and the Earth have been created.
>
> C. S. PEIRCE

1. *Introduction.* If U_1, U_2, ..., and V are sentences (which may be either ue or false), then we often say that V is (or is not) a LOGICAL CONSEQUENCE U_1, U_2, ..., or that the conclusion V LOGICALLY FOLLOWS from the emisses U_1, U_2, ...; for instance, we might say that the conclusion:

Some Panthers are not Swans,

GICALLY FOLLOWS from the premisses:

Some Panthers are not Mammals,
Some Mammals are not Swans.

It is the task of logic (and, in fact, its main task) to clarify this notion and point out the conditions for its application. Now if we try to carry out is task, we find that the term '*logical consequence*' covers two distinct tions which, approximately, can be characterized as follows.

(i) *Formal Derivability.* There are certain formal *rules of inference*, each which, if applied to appropriate premisses, yields a certain *immediate nclusion.* Well-known rules of inference are, for instance, the *modus nens*:

$$\frac{\textit{If } P \textit{ then } Q;\quad P}{Q}$$

d the *conversio simplex*:

$$\frac{\textit{Some } A\textit{'s are } B;}{\textit{Some } B\textit{'s are } A.}$$

rom *Mededelingen van de Koninklijke Nederlandse Akademie van Wetenschappen, leling Letterkunde*, N.R. Vol. 18, no. 13 (Amsterdam, 1955), pp. 309–42. Reprinted permission of the Royal Netherlands Academy of Sciences.

Now V is called a LOGICAL CONSEQUENCE of U_1, U_2, ..., if it is *formall derivable* from U_1, U_2, ..., that is, if, starting from the premisses U_1, U_2, ... and applying again and again the rules of inference, we can finally obtai the conclusion V.

(The rules of inference are called '*formal*' on account of the fact that the can be stated in purely '*typographical*' terms, without any reference to th meaning of the sentences to which they are applied.)

(ii) *Semantic Entailment*. It is known from Aristotle's syllogistics that th conclusion: *Some Panthers are not Swans*, does *not* LOGICALLY FOLLOW fror the premisses: *Some Panthers are not Mammals*, and: *Some Mammals ar not Swans*. If we wish to show this, we usually argue as follows. Let u replace the terms '*Panther*', '*Swan*', and '*Mammal*' by '*Pig*', '*Swine*', an '*Mammoth*', respectively. Then we get *new* premisses, namely:

Some Pigs are not Mammoths,
Some Mammoths are not Swine,

and a *new* conclusion, namely:

Some Pigs are not Swine.

Now the new premisses are true, whereas the new conclusion is fals Hence the new conclusion cannot LOGICALLY FOLLOW from the ne premisses and, in view of the similarity as to logical form, the old conclusio does not LOGICALLY FOLLOW from the old premisses.

In this discussion, the '*truth value*' (that is, the truth or falsehood) of th new premisses and conclusion and hence also the *meaning* of the terms '*Pig* '*Swine*', and '*Mammoth*' play an essential role; therefore, that notion LOGICAL CONSEQUENCE which is relevant in this context may be denoted *semantic entailment*. It can be defined as follows: V is said to be *semantical entailed* by U_1, U_2, ..., if we cannot replace the terms in U_1, U_2, ..., and by new terms in such a manner that the new premisses U_1^*, U_2^*, ... are tr whereas the new conclusion V^* is false.

If, however, as in the above example, such new terms *can* be found (which case V is *not* semantically entailed by U_1, U_2, ...), then we say th these terms provide us with a suitable *counter-example* for proving that does *not* LOGICALLY FOLLOW from U_1, U_2, ...

It will be clear that, on the other hand, the meaning of the old terr '*Panther*', '*Swan*', and '*Mammal*' is completely irrelevant; in fact, the abo counter-example shows that no conclusion of the form:

Some P's are not S,

is a LOGICAL CONSEQUENCE of the premisses:

Some P's are not M,
Some M's are not S.

Such expressions, containing *indeterminates* '*P*', '*M*', '*S*', which stand for arbitrary terms, were already frequently used in traditional logic. Modern logic instead uses formulas:

$$(Ez)\ [P(z)\ \ \&\ \overline{S(z)}],$$
$$(Ex)\ [P(x)\ \ \&\ \overline{M(x)}],$$
$$(Ey)\ [M(y)\ \&\ \overline{S(y)}].$$

Though the meaning of such formulas will be pretty clear from the present context, it will be better to say a few words about the manner in which thay are to be interpreted.[1]

We select some (non-empty) set of individuals, which will be called the *universe of discourse* (it may, for instance, contain (i) all human beings, or (ij) all natural numbers). The letters '*P*', '*M*', '*S*', ... stand for terms (properties or predicates) which can be applied to some individuals in the universe (for instance, (i) *healthy*, *male*, *grown-up*, or (ij) *odd*, *prime*, *square*). The letters '*x*', '*y*', '*z*', ... are used as *variables* 'ranging' over the universe. The *atomic formula* '$P(x)$' is used to state the condition for a certain individual x to have the property P. The *quantifier* '(x)' is used to express the fact (or the supposition) that *all* individuals in the universe fulfil the condition which follows the quantifier; the quantifier '(Ex)' is used to express the fact (or the supposition) that *some* individuals in the universe fulfil the condition by which it is followed. The quantifiers '(y)', '(z)', ..., '(Ey)', '(Ez)', ... are to be interpreted in the same manner. The symbols:

$$\overline{\quad},\quad \&,\quad \rightarrow$$

are read: '*not*', '*and*', and '*if* ..., *then*'. Later on, we also use the symbols:

$$\vee,\quad \leftrightarrow$$

which are read: '*or*' and '*if and only if*'.

A *counter-example* (which may or may not be suitable in some context) is obtained by selecting a universe and by selecting properties **P**, **M**, **S**, ... for all letters '*P*', '*M*', '*S*', ...[2]

[1] A more detailed exposition is found, for instance, in my *Fondements logiques des mathématiques* (2me éd., Paris, 1955).

[2] It is, perhaps, more customary to use the term '*model*' or '*interpretation*' for what is here called a 'counter-example'. Usually, a model or an interpretation is only

2. *The Completeness Theorem.* The two notions of LOGICAL CONSEQUENCE which have been described in Section 1 are both used in applied logic; this fact creates a situation which, from a methodological point of view, is far from satisfactory. I wish first to explain the situation and then to discuss its implications; my explanation is based upon the consideration that in some contexts the notion of formal derivability is more helpful, whereas in other contexts we naturally tend to apply the notion of semantic entailment.

(i) Suppose that we wish to show that V is a LOGICAL CONSEQUENCE of $U_1, U_2, \ldots$; then it is natural to resort to the notion of formal derivability. We try to establish a *formal derivation* of V from $U_1, U_2, \ldots$, that is, a sequence of applications of the rules of inference which, starting from the premisses $U_1, U_2, \ldots$, finally yields the conclusion V; it is obviously sufficient to point out *one* such derivation.

If instead we should wish to apply the notion of semantic entailment, then it would be necessary to examine *all* possible counter-examples and to show that among them *no* suitable counter-example for proving that V is *not* a LOGICAL CONSEQUENCE of $U_1, U_2, \ldots$ can be found.

(ij) Suppose, on the other hand, that we wish to show that V does *not* LOGICALLY FOLLOW from $U_1, U_2, \ldots$; then we would prefer to apply the notion of semantic entailment. For it is sufficient to point out *one* suitable counter-example for proving that V does not LOGICALLY FOLLOW from $U_1, U_2, \ldots$. But if we should wish to apply the notion of formal derivability, then it would be necessary to examine *all* formal derivations which start from the premisses $U_1, U_2, \ldots$ and to point out that *none* of them yields the conclusion V.

However, the practical advantages which can be obtained by using simultaneously two different notions of LOGICAL CONSEQUENCE cannot, of course, justify such a procedure. Specifically, by accepting it we incur the risk of obtaining conflicting results.

(i) If the notion of formal derivability is wider than the notion of semantic entailment, then it may happen that a conclusion V is formally derivable from certain premisses $U_1, U_2, \ldots$ without being semantically entailed by them; if V is not semantically entailed by $U_1, U_2, \ldots$, then there is some suitable counter-example for proving this fact. This counter-example enables us to replace $U_1, U_2, \ldots, V$ by new sentences $U_1^*, U_2^*, \ldots, V^*$ such that V^* is again formally derivable from $U_1^*, U_2^*, \ldots$, whereas $U_1^*, U_2^*, \ldots$

denoted as a 'counter-example', if it is suitably chosen in view of refuting a certain supposition. But I feel that in the present context the above terminology is more suggestive. The traditional name for a counter-example is '*instance*' (ἔνστασις, *instantia*; cf. Aristotle, *Prior Analytics* B 26, 69ª 37). It is not necessary here to discuss the rather obvious connections with the *parable.*

are *true* and V^* is *false*. Now in the formal derivation of V^* from U_1^*, U_2^*, ..., there must occur some rule of inference which is applied to certain true premisses X, X', ... and yields an immediate conclusion Y which is false.

(ij) Suppose the notion of semantic entailment to be wider than the notion of formal derivability. Then it may happen that a conclusion V is semantically entailed by the premisses U_1, U_2, ... but not formally derivable from them. In other words: V is not formally derivable from U_1, U_2, ..., but there is no suitable counter-example for proving that it does not LOGICALLY FOLLOW from them.

In both cases we throw the blame for the discrepancy upon the notion of formal derivability (which shows, by the way, that the notion of semantic entailment is the more fundamental one of the two) and we try to regularize the situation by revising the rules of inference. If this attempt is successful, then the formal rules of inference (i) do not yield a false conclusion if applied to true premisses, and (ij) enable us to derive the conclusion V from the premisses U_1, U_2, ... whenever it is semantically entailed by them. Now the two notions of LOGICAL CONSEQUENCE are *equipollent* (they have the same extension) and hence they may be used indifferently.

In the systematic construction of a logical theory, one usually proceeds as follows.

(I) The rules of inference are stated and the notion of formal derivability is introduced.

(II) The notion of truth (or a related notion) is introduced, and it is shown that from true premisses no false conclusion can be formally derived.

(III) The notion of semantic entailment is introduced, and it is shown that, whenever V is semantically entailed by U_1, U_2, ..., it can also be formally derived from them. This part of the construction, which is known as the proof of the Completeness Theorem[3] for the logical theory under consideration, usually presents considerable difficulties.

On the basis of investigations, which I have made during the last few years,[4] I am now able to construct logical theories in a different manner so as to avoid the above difficulties. The main idea is, that these difficulties derive from the fact that, in looking for suitable counter-examples, we did not proceed in a systematic manner.

[3] The first proof of this kind has been given by Kurt Gödel, 'Die Vollständigkeit der Axiome des logischen Funktionenkalküls', *Monatsh. Math. Phys.*, Vol. 37 (1930).

[4] E. W. Beth, 'A Topological Proof of the Theorem of Löwenheim–Skolem–Gödel', *Proceedings*, Vol. 54 (1951); 'Some Consequences of the Theorem of Löwenheim–Skolem–Malcev', ibid., Vol. 56 (1953); 'On Peano's Method in the Theory of Definition', ibid.; 'Remarks on Natural Deduction', ibid. Vol. 58 (1955).
I shall later refer to related work by other authors.

3. *Heuristic Considerations: Construction of Semantic Tableaux.* Let us consider two concrete problems, namely:

(i) Is the formula: $(Ez)\,[P(z)\ \&\ \overline{S(z)}]$ a logical consequence of the formulas: $(Ex)\,[(P(x)\ \&\ \overline{M(x)}]$ and $(Ey)\,[M(y)\ \&\ \overline{S(y)}]$?

(ij) Is the formula: $(Ez)\,[S(z)\ \&\ \overline{P(z)}]$ a logical consequence of the formulas: $(x)\,[P(x) \rightarrow \overline{M(x)}]$ and $(Ey)\,[S(y)\ \&\ M(y)]$?

We interpret the notion of LOGICAL CONSEQUENCE as semantic entailment.

In order to solve such a problem, we try to show, by pointing out a suitable counter-example, that the first formula does *not* logically follow from the second and third ones. If such a counter-example is found, then we have a negative answer to our problem. And if it turns out that no suitable counter-example can be found, then we have an affirmative answer. In this case, however, we must be sure that no suitable counter-example whatsoever is available; therefore, we ought not to look for a counter-example in a haphazard manner, but we must rather try to construct one in a systematic way. Now there is indeed a systematic method for constructing a counter-example, if available; it consists in drawing up a *semantic tableau.* Let us consider the tableaux which correspond to the above problems (i) and (ij), respectively.

ad (i) The tableau makes it clear that in this case we can indeed find a suitable counter-example; it follows that (in agreement with Aristotle's syllogistics, and with the result of our discussion in Section 1) the first formula does *not* logically follow from the second and third ones.

Valid		*Invalid*	
(1) $(Ex)\,[P(x)\ \&\ \overline{M(x)}]$		(3) $(Ex)\,[P(z)\ \&\ \overline{S(z)}]$	
(2) $(Ey)\,[M(y)\ \&\ \overline{S(y)}]$		(7) $M(a)$	(6)
(4) $P(a)\ \&\ \overline{M(a)}$	(1)	(8) $P(a)\ \&\ \overline{S(a)}$	(3)
(5) $P(a)$		(9) $\overline{S(a)}$	(5) and (8)
(6) $\overline{M(a)}$		(14) $S(b)$	(13)
(10) $S(a)$	(9)	(15) $P(b)\ \&\ \overline{S(b)}$	(3)
(11) $M(b)\ \&\ \overline{S(b)}$	(2)	(16) $P(b)$	(13) and (15)
(12) $M(b)$			
(13) $\overline{S(b)}$			

In constructing the above tableau, we proceed as follows, The lines (1)–(3) simply state the conditions to be satisfied by any suitable counter-example. On account of line (1), there must be some individual which fulfils the condition: $P(x)$ & $\overline{M(x)}$; *let this individual be given the name* 'a'. Then we have line (4) and hence lines (5)–(7). By line (3), the individual a must *not* fulfil the condition $P(z)$ & $\overline{S(z)}$; this accounts for lines (8)–(10). Likewise, by line (2), there must be some individual which fulfils the condition: $M(y)$ & $\overline{S(y)}$; *if we give it the name* 'b', then we obtain line (11) and hence lines (12)–(16).

As we have taken care of all conditions (1)–(3), there is no reason to introduce additional individuals; so we may consider the construction to be successfully terminated. We observe that, to each of the atoms $P(a)$, $P(b)$, $M(a)$, $M(b)$, $S(a)$, and $S(b)$, the tableau assigns a definite '*truth value*'. So the counter-example provided by the tableau can be described as follows. The universe consists of two individuals, called 'a' and 'b'; the property **P** belongs to a but not to b, and so does the property **S**; the property **M**, on the other hand, belongs to b but not to a.

A suitable counter-example in connection with problem (i) may, of course, involve other individuals besides those called 'a' and 'b'. One such counter-example was already mentioned in Section 1. Another one is shown

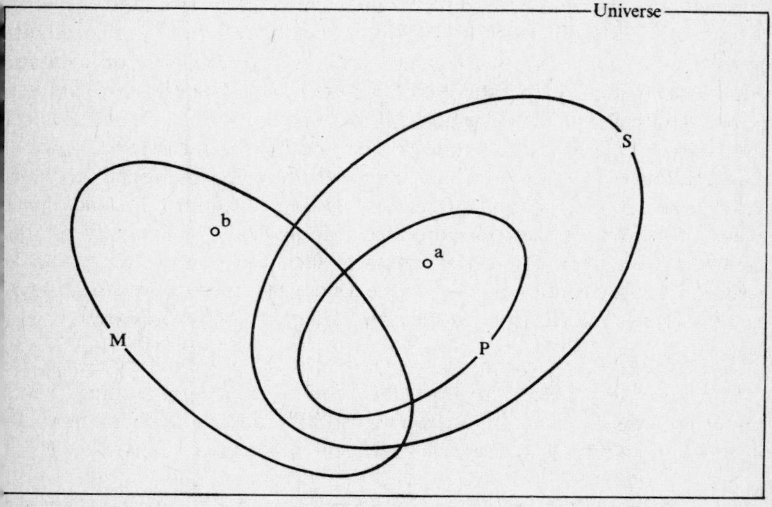

Fig. 1

by Fig. 1, where the notions **P**, **M**, and **S** are represented by regions (such diagrams were frequently used in traditional logic).

ad (ij) In this case, our tableau tells a different story: our systematic attempt (and, in fact, any attempt) at constructing a suitable counter-example for proving, that the first formula does *not* logically follow from the second and third one, breaks down.

Valid		*Invalid*	
(1) $(x)\,[P(x) \to \overline{M(x)}]$		(3) $(Ez)\,[S(z)\ \&\ \overline{P(z)}]$	
(2) $(Ey)\,[S(y)\ \&\ M(y)]$		(8) $S(a)\ \&\ \overline{P(a)}$	
(4) $S(a)\ \&\ M(a)$		(9) $\overline{P(a)}$	(5) and (8)
(5) $S(a)$		(12) $M(a)$	(11)
(6) $M(a)$	(4)		
(7) $P(a) \to \overline{M(a)}$	(1)		
(10) $P(a)$	(9)		
(11) $\overline{M(a)}$	(7) and (10)		

The failure of our systematic attempt appears on line (12), where we are compelled to assign to $M(a)$ a truth value which differs from the one which it had previously obtained on line (6). From the failure of our systematic attempt it follows that any other attempt must fail as well. For a suitable counter-example has to satisfy conditions (1)–(3). Therefore, its universe must contain an individual which fulfils the formula: $S(y)\ \&\ M(y)$; and if this individual is given the name 'a', then we obtain line (4) and hence all the remaining lines in the tableau. So an arbitrary counter-example which satisfies conditions (1) and (2) cannot satisfy condition (3); and this is exactly what we express by saying that formula (3) is semantically entailed by formulas (1) and (2). As a matter of fact, this result corresponds to Aristotle's rule for the syllogism in the modus FESTINO; and even our argument is closely related to Aristotle's method of proof by ἔκθεσις.[5]

4. *Heuristic Considerations: Transformation of Semantic Tableaux.* Let us rearrange our tableau for problem (ij) in the following manner: we do not change the left column, but we extend it by adding below all formul in the right column in the reverse order; the result is as follows.

[5] I. M. Bocheński, *Ancient Formal Logic* (Amsterdam, 1951), p. 47; J. Lukasiewicz *Aristotle's Syllogistic* (Oxford, 1951), pp. 59 ff.

(1)	$(x)\,[P(x) \rightarrow \overline{M(x)}]$	(prem)
(2)	$(Ey)\,[S(y)\ \&\ M(y)]$	(prem)
(4)	$S(a)\ \&\ M(a)$	(+hyp 1)
(5)	$S(a)$	(4)
(6)	$M(a)$	(4)
(7)	$P(a) \rightarrow \overline{M(a)}$	(1)
(10)	$P(a)$	(+hyp 2)
(11)	$\overline{M(a)}$	(7) and (10)
(9)	$\overline{P(a)}$	(−hyp 2)
(8)	$S(a)\ \&\ \overline{P(a)}$	(5) and (9)
(3)	$(Ez)\,[S(z)\ \&\ \overline{P(z)}]$	(−hyp 1)

This ought to be a pleasant surprise! For it provides us with a formal lerivation, made up of step-by-step inferences, each of which is an appliation of some formal rule of inference; let us consider the argument in nore detail.

On line (4) we introduce, besides the 'given premisses' (1) and (2), an dditional hypothesis. Part of the derivation is made 'under the hypothesis '; this is indicated by *single* horizontal lines at the beginning and at the end. The conclusions (5)–(7) are self-explanatory.

On line (10), a second hypothesis is introduced. The part of the deriation which is made 'under the hypothesis 2' is indicated by *double* orizontal lines. On line (11), we apply *modus ponens*. Hypothesis 2 thus urns out to entail a formal contradiction; the appropriate conclusion *x absurdo* is drawn on line (9); the remaining part of the derivation no onger depends upon hypothesis 2. The conclusion on line (8) is again selfxplanatory.

And now follows, on line (3), another important step. It might appear as f the statement on this line, namely, $(Ez)\,[S(z)\ \&\ \overline{P(z)}]$, ought still to be nade 'under the hypothesis' that we have $S(a)\ \&\ M(a)$. For the context uggests that in the derivation of formula (3) the special choice of the inlividual a might somehow play a certain role. However, this cannot be the ase, as neither in the premisses (1) and (2) nor in the conclusion (3) the ndividual a is even mentioned. Hence any other individual, say y, would erve our purpose equally well, provided it fulfils the condition $S(y)\ \&$ $M(y)$. It follows that the conclusion (3) can be drawn, if some element y f this kind is available; but this is indeed the case on account of the remiss (2). So hypothesis 1 is now superfluous, and this idea is expressed on ine (3).

The derivation which has been discussed presents a few outstanding characteristics, on which I should like to lay some stress.

(i) It is in complete harmony with the standpoint of semantics, as it has been obtained by a transformation of a tableau which we had established on the basis of semantic considerations concerning the *meaning* of the words '*all*', '*some*', '*not*', '*and*', and '*if* ..., *then*'.

(ij) Nevertheless, and in full agreement with the point of view of a formal theory of derivability, it can alternatively be construed as a sequence of applications of certain rules of inference, which can be formulated in purely 'typographical' terms, without any reference to the meaning of the sentences (or formulas) to which these rules are applied.

(iij) It follows very closely the manner in which we argue if we do not try to adapt ourselves to some preconceived logical theory. A systematization of the principles, underlying our derivation, and all similar ones, would result in the construction of a certain *System of Natural Deduction* as given, first by G. Gentzen, and later by various other authors.[6] Curiously enough, these authors were guided in their constructions by considerations pertaining to the formal theory of derivability rather than by semantic conceptions. R. Carnap, K. R. Popper, W. V. Quine, and H. Scholz constructed the theory of derivability on the basis of semantic considerations, but these authors did not establish those close connections between semantics and derivability theory which I shall try to point out.[7]

(iv) It automatically satisfies Gentzen's *subformula principle*, as the construction of the original tableau consisted in dissolving both the premisses and the conclusion into ever smaller '*subformulas*'. Therefore, such celebrated and profound results as Herbrand's Theorem, the Theorem of Löwenheim–Skolem–Gödel, Gentzen's Subformula Theorem and Extended Haupsatz, and Bernays' Consistency Theorem are, from our present standpoint, within (relatively) easy reach. At the same time, our approach

[6] G. Gentzen, 'Untersuchungen über das logische Schliessen'. *Math. Zeitschr.*, Vol 39 (1934); H. B. Curry, *A Theory of Formal Deducibility* (Notre Dame, Indiana, 1950) K. Schütte, 'Schlussweisen-Kalküle der Prädikatenlogik', *Math. Ann.*, Vol. 122 (1950) G. Hasenjaeger, 'Konsequenzenlogik', *in*: H. Hermes–H. Scholz, *Mathematische Logik*, Enz. d. math. Wiss. I 1, 1, i (Berlin, 1952); S. C. Kleene, *Introduction to Metamathematics* (Amsterdam-Groningen, 1952).

An expository treatment of research in this direction is found in: R. Feys, 'Les méthodes récentes de déduction naturelle', *Revue Philos. de Louvain*, Vol. 44, (1946) 'Note complémentaire sur les méthodes de déduction naturelle', ibid. Vol. 45 (1947) *Logistiek*, deel 1 (Antwerpen-Amsterdam, 1945).

[7] R. Carnap, *Studies in Semantics*, Vol. I: Introduction to Semantics, Vol. II: Formalization of Logic (Cambridge, Mass., 1942–3); K. R. Popper, 'New Foundations for Logic', *Mind*, Vol. 56 (1947); Errata, ibid. Vol. 57 (1948); W. V. Quine, *Methods of Logic* (New York, 1950); H. Scholz, *Vorlesungen über Grundzüge der mathematischen Logik*, 2 Bde (Münster i/Westf., 1950–1).

ealises to a considerable extent the conception of a purely *analytical method*, vhich has played such an important role in the history of logic and philsophy.[8]

5. *Outline of Systematic Treatment*. As a result of our previous discussion, he prospects for a simple and elegant treatment of the problems of logic eem to be very hopeful. Let us adopt the above problems (i) and (ij) as aradigmatic cases. Then it seems that with respect to any problems of this ind (the treatment of which provides the starting-point, and in fact the very ubstance, of logical investigation) the situation can be described as follows. he problem of finding out, whether or not a certain conclusion V LOGIALLY FOLLOWS from given premisses U_1, U_2, ..., can be treated by two ifferent methods which correspond to two different notions of LOGICAL ONSEQUENCE.

(i) We try to find a straightforward derivation of V from U_1, U_2, ..., by pplying the rules of inference of a certain System of Natural Deduction F. ccording as such a derivation can or cannot be found, the answer to our roblem is affirmative or negative.

(ij) We try to construct a suitable counter-example for proving that V is *ot* a logical consequence of U_1, U_2, ... According as our systematic attempt s or is not successful, the answer to our problem is negative or affirmative.

In accordance with the demand which was formulated in Section 2, these wo methods are equivalent (hence for the System F the Completeness heorem holds true). For if the semantic tableau shows that the attempt at onstructing a suitable counter-example breaks down, then we can rearrange he tableau so as to obtain a straightforward derivation in the System F; nd, conversely, if we have a derivation in the System F, it follows that any ttempt at constructing a suitable counter-example must break down (for he System F does not allow the derivation of a false conclusion from true remisses).

Fundamentally, this description of the situation is correct; however, onsiderable caution is in order, as the situation is more involved than one night infer. For so far our discussion has been rather fragmentary; a ystematic treatment, culminating in a proof of the completeness theorem or the System F, demands the following steps.

(I) We must state precise rules for the construction of semantic tableaux.

(II) We must give a precise description of the System F.

(III) We must state precise rules for the transformation of a semantic ableau into a formal derivation, and vice versa.

[8] Plato, *Philebus* 18 B-D; Aristotle, *Metaphysics* Γ 3, 1005^b 2; Leibniz, *Opuscules et agments inédits de* —, extraits des manuscrits par L. Couturat (Paris, 1903), *passim*.

(IV) It may happen that the (tentative) construction of a suitable counter-example demands infinitely many steps. We have to show that in such cases the construction (or the corresponding semantic tableau) provides us with a suitable counter-example. In doing so, we have to take into account the possibility of still other complications which did not present themselves in our examples.

6. *Rules for Construction and Transformation of Semantic Tableaux.* It may happen that a tableau must be split up. In this case, we obtain two left *subcolumns*, each of which is considered as a continuation of the original left column; this means that each of these two left subcolumns is supposed to contain all formulas which are contained in the original left column; we also obtain two right subcolumns, each of which is a continuation of the original right column. The subcolumns are pairwise combined in such a manner as to form two *subtableaux*; a left and a right column (or subcolumn) which together form one tableau (or subtableau) are said to be *conjugate*. A subtableau can again be split up, and so on; a subtableau which has been obtained by the (possibly repeated) splitting up of a given tableau (or subtableau) is said to be *subordinate* to that tableau (or subtableau). In a wider sense the term '*tableau*' will also be used to denote a tableau (in the narrower sense) together with its subordinate subtableaux.

Now the rules for the construction of a semantic tableau can be stated as follows.

(i) In the left column of the original tableau we may insert arbitrary *initial formulas* $U_1, U_2, \ldots$; in its right column likewise arbitrary initial formulas $V_1, V_2, \ldots$.

(ij) If $\bar{X}$ appears in some column, then we insert X in the conjugate column.

(iij) If $(x)X(x)$ appears in a left column or $(Ex)X(x)$ in a right column, then we insert in the same column $X(p)$ for each individual p which has been or will be introduced.

(iv) If $X \,\&\, Y$ appears in a left column or $X \vee Y$ in a right column, then we insert both X and Y in the same column.

(v) If $X \rightarrow Y$ appears in a right column, then we insert Y in the same column and X in the conjugate left column.

(vi) If (a) $X \vee Y$ appears in a left column, (b) $X \rightarrow Y$ in a left column, (c) $X \,\&\, Y$ in a right column, (d) $X \leftrightarrow Y$ in a left column, or (e) $X \leftrightarrow Y$ in a right column, then the tableau must be split up, and we insert:

(a) X in one left subcolumn and Y in the other.

(b) Y in one left column and X in the right column *not* conjugate with it.

(c) X in one right subcolumn and Y in the other.

(d) Both X and Y in one left column and once more both X and Y in the right column *not* conjugate with it.

(e) X in one left subcolumn, Y in the right subcolumn conjugate with it, Y in the other left subcolumn, and X in the remaining right subcolumn.

(vij) If $(Ex)\,X(x)$ appears in a left column or $(x)\,X(x)$ in a right column, en we introduce a new individual p and we insert $X(p)$ in the same column : is convenient not to apply this rule until all possibilities of applying other les have been exhausted; the part of the construction which starts with the troduction of the kth individual and which terminates at the introduction the $(k+1)^{\text{st}}$ one is denoted as its kth *stage*).

(viij) If one and the same formula appears in two conjugate columns, en the corresponding tableau (or subtableau) is closed.

(ix) If all subtableaux subordinate to some tableau (or subtableau) are osed, then that tableau (or subtableau) itself is also closed.

(x) If all possibilities of applying rules (i)–(vij) have been exhausted, hereas the tableau is not closed, then the tableau provides us with a suitable unter-example, and is terminated.

It will be clear that the rules (i)–(vij) exactly reflect the *meaning* of the ords '*all*' '*some*', '*not*', '*and*', '*if ..., then*', '*or*', '*if and only if*', and of the mbols by which they have been replaced. For instance, part of rule (iv) rresponds to the following rule of semantics: The sentence: 'X *and* Y' is ue (and the formula: 'X & Y' is valid), if and only if both X and Y are true r valid). However, these rules assume a completely formal character, if stead of speaking about an individual p we speak about a symbol 'p'.

Perhaps it will be useful to give an example of the application of the rules. et us construct a semantic tableau, taking as initial formulas:

$$(x)(y)\,[A(x) \vee B(y)]$$

the left column, and:

$$(x)\,A(x) \vee (y)\,B(y)$$

the right column (see tableau on p. 22).

We have now taken care of point (I) on the programme outlined in ction 5. So we turn to points (II) and (III); these points can be handled ry quickly, by stating the following definitions:

(i) A formal proof in the System of Natural Deduction F of the *sequent*:

$$U_1, U_2, \ldots \vdash V_1, V_2, \ldots$$

a closed semantic tableau in which the initial formulas are: U_1, U_2, ... in e left column, and V_1, V_2, ... in the right column.

<table>
<tr><th colspan="3">Valid</th><th colspan="3">Invalid</th></tr>
<tr><td colspan="3">(1) $(x)(y)\,[A(x) \vee B(y)]$
(7) $(y)\,[A(a) \vee B(y)]$
(8) $A(a) \vee B(a)$</td><td colspan="3">(2) $(x)\,A(x) \vee (y)\,B(y)$
(3) $(x)\,A(x)$
(4) $A(a)$
(5) $(y)\,B(y)$
(6) $B(b)$</td></tr>
<tr><td>(i)
(9) $A(a)$</td><td colspan="2">(ij)
(10) $B(a)$
(11) $(y)\,[A(b) \vee B(y)]$
(12) $A(a) \vee B(b)$</td><td>(i)</td><td colspan="2">(ij)</td></tr>
<tr><td></td><td>(iij)
(13) $A(a)$</td><td>(iv)
(14) $B(b)$</td><td></td><td>(iij)</td><td>(iv)</td></tr>
</table>

(ij) A formal derivation in F of the conclusion V from the premiss $U_1, U_2, \ldots$ is a formal proof in F of the sequent:

$$U_1, U_2, \ldots \vdash V.$$

(iij) A formal derivation in F of the formula U as a *logical contradicti* is a formal proof in F of the sequent:

$$U \vdash \emptyset$$

(the symbol '∅' is used to express the fact that no formula appears at t right side of the sequent).

(iv) A formal derivation in F of the formula V as a *logical identity* is formal proof in F of the sequent:

$$\emptyset \vdash V.$$

The main content of these definitions could also be stated as follows: formal derivation in F of the conclusion V from the premisses $U_1, U_2, \ldots$ a closed semantic tableau in which the initial formulas are: $U_1, U_2, \ldots$ in t left column, and V in the right column.

It will be clear that on account of this treatment of point (III) we avo the necessity of dealing with point (IV). It can, however, be argued that it preferable to give the derivations in F a more familiar shape.

We shall not yield to this argument, but nevertheless it is interesting show that, if necessary, we could easily give the formal derivations in F more normal appearance. Specifically, we could either borrow the gener

ucture of Gentzen's Calculus of Natural Deduction or adapt the prin-
les underlying his Calculus of Sequents.

Though the connections of our semantic tableau to some System of tural Deduction were already discussed in Section 4, it will be useful also give our above tableau the familiar shape of a derivation in a system of s kind.

$$\begin{array}{rll}
(1) & (x)(y)\,[A(x) \vee B(y)] & \text{(prem)} \\
(7) & (y)\,[A(a) \vee B(y)] & \\
(12) & A(a) \vee B(b) & \text{(dilemma)} \\ \cline{2-2}
(6) & B(b) & (+\text{alt } 1) \\
(5) & (y)\,B(y) & \\ \cline{2-2}
(4) & A(a) & (+\text{alt } 2) \\
(3) & (x)\,A(x) & \\ \cline{2-2}
(2) & (x)\,A(x) \vee (y)\,B(y) & (-\text{alt } 1, 2)
\end{array}$$

f we try to transform the closed tableau into a proof in some Calculus of quents, the result is as follows. We obtain a version of this Calculus, ich is closely related to Hasenjaeger's Symmetric Calculus.[6]

$$\begin{array}{ll}
A(a) \vdash A(a) & B(b) \vdash B(b) \\
A(a) \vdash A(a), B(b) & B(b) \vdash A(a), B(b) \\ \hline
\multicolumn{2}{c}{A(a) \vee B(b) \vdash A(a), B(b)} \\
\multicolumn{2}{l}{(x)(y)\,[A(x) \vee B(y)] \vdash A(a), B(b)} \\
\multicolumn{2}{l}{(x)(y)\,[A(x) \vee B(y)] \vdash (a)\,A(x), (y)\,B(y)} \\
\multicolumn{2}{l}{(x)(y)\,[A(x) \vee B(y)] \vdash (x)\,A(x) \vee (y)\,B(y)}
\end{array}$$

As these formal derivations have been constructed, so to speak, in a rely mechanical manner (we have indeed come alarmingly near the lization of the 'ideal' of a *calculus ratiocinator*, a 'logical machine'; I ll return to this point later on), one might expect them to be rather msy and cumbersome. But this anticipation is not corroborated; to the ntrary, the above derivations are remarkably concise, and the derivations tained by our procedure can even be proved to be, in a sense, the shortest es which are possible. The first derivation, for instance, uses a rather ring *dilemma*, which I would certainly have avoided (by means of a *uctio*), but which is, in view of the tableau, completely justified.

f we are submitted a derivation (belonging to F or to some other system, h as those in Hilbert-Ackermann's *Grundzüge* or in Quine's *Methods of gic*) which seems to be incorrect or needlessly involved, then, by con-ucting a suitable tableau, we obtain a correct or a simplified derivation, ovided such a derivation can be found. In particular, an indirect proof will be replaced by a direct one, whenever such a proof is available; this an interesting result in connection with a paper on this subject by Löwe heim.[9]

7. *Infinite Tableaux*. I shall now show that in certain cases the constru tion of a semantic tableau involves infinitely many steps, and discuss th possibility of pointing out the existence of a suitable counter-example such a situation arises.

(1) The first, and rather obvious, case is that of a sequent:

$$U_1, U_2, \ldots, \textit{ad inf.} \vdash V,$$

which involves infinitely many premisses U_1, U_2, ... In this case, it is co venient slightly to change the division of the construction into successi stages, as follows. We begin the first stage by taking into account the initi formulas U_1 and V and by introducing one individual a. We pass on to new stage either by taking into account a new formula U_j or by introduci a new individual p; if the kth stage started with the introduction of a ne individual p, then the $(k+1)^{\text{st}}$ stage *must* start with taking into account new initial formula U_j; and if the kth stage started with taking into accou a new initial formula U_j, then the $(k+1)^{\text{st}}$ stage will *normally* start with th introduction of a new individual p; however, it may happen that at the en of the kth stage we have no reason to introduce a new individual, and in th case the $(k+1)^{\text{st}}$ stage will start with taking into account still another ne initial formula U_{j+1}.—It will be clear that the above sequent will be provab in F, if and only if, for some j, the sequent:

$$U_1, U_2, \ldots, U_j \vdash V$$

is provable in F. Accordingly, the conclusion V is derivable in F from th premisses U_1, U_2, ..., *ad inf.*, if and only if, for some j, it is derivable fro the premisses U_1, U_2, ..., U_j.

(2) Even one single initial formula U may give rise to an infinite tablea provided it contains some *binary predicate* (or *relation*). Let us consider an example the sequent:

$$U \vdash \emptyset,$$

where U is a certain formula studied by K. Schütte (1934):

$$(x)\,[\overline{R(x,x)}\ \&\ (Ey)\,\{R(x,y)\ \&\ (z)\,[R(z,x) \rightarrow R(z,y)]\}].$$

9 L. Löwenheim, 'On Making Indirect Proofs Direct', translated by W. V. Qui *Scripta Math.*, Vol. 12 (1946); the author refers to earlier work by Bolzano23 a G. Hessenberg. Cf. R. L. Goodstein, 'Proof by *reductio ad absurdum*', *Math. Gaze* Vol. 32 (1948).

Valid		*Invalid*
(1) $(x)\,[\overline{R(x,x)}\,\&\,(Ey)\{R(x,y)\,\&\,(z)\,[\ldots]\}]$		
irst stage		(4) $R(1,1)$
(2) $\overline{R(1,1)}\,\&\,(Ey)\{\ldots\}$	(3) $\overline{R(1,1)}$	
(5) $(Ey)\,\{R(1,y)\,\&\,(z)\,[\ldots]\}$		
nd stage		
(6) $R(1,2)\,\&\,(z)\,[\ldots]$	(7) $R(1,2)$	
(8) $(z)\,[R(z,1) \rightarrow R(z,2)]$		(10) $R(1,1)$
(9) $R(1,1) \rightarrow R(1,2)$		(13) $R(2,1)$
	(11) $R(1,2)$	
(12) $R(2,1) \rightarrow R(2,2)$		(17) $R(2,2)$
	(14) $R(2,2)$	
(15) $\overline{R(2,2)}\,\&\,(Ey)\{\ldots)$	(16) $\overline{R(2,2)}$	
(18) $(Ey)\{R(2,y)\,\&\,(z)[\ldots]\}$		
hird stage		
(19) $R(2,3)\,\&\,(z)[\ldots]$	(20) $R(2,3)$	
(21) $(z)\,[R(z,2) \rightarrow R(z,3)]$		(29) $R(3,2)$
(22) $R(1,2) \rightarrow R(1,3)$	(24) $R(1,3)$	
(25) $R(2,2) \rightarrow R(2,3)$		(32) $R(3,1)$
(28) $R(3,2) \rightarrow R(3,3)$		
(31) $R(3,1) \rightarrow R(3,2)$	cf. (8)	(36) $R(3,3)$
—		—
th stage		
$R(k-1,k), R(1,k), \ldots, R(k-2,k)$		$R(k,k-1), R(k,1), \ldots,$
$\overline{R(k,k)}\,\&\,(Ey)\,\{\ldots\}$		$R(k,k-2), R(k,k)$
$+1)^{st}$ stage		
$R(k,k+1)\,\&\,(z)\,[\ldots]$	$R(k,k+1)$	
$R(1,k) \rightarrow R(1,k+1), \ldots$	$R(1,k+1), \ldots$	
$R(k-1,k) \rightarrow R(k-1\ \ k+1)$	$R(k-1,k+1)$	
$R(k,k) \rightarrow R(k,k+1)$		
$R(k+1,k) \rightarrow R(k+1,k+1)$		$R(k+1,k)$
$R(k+1,1) \rightarrow R(k+1,2)$	cf. (8)	$R(k+1,1)$
$\ldots, R(k+1,k-1) \rightarrow R(k+1,k)$		$\ldots, R(k+1,k-1)$
$\overline{R(k+1,k+1)}\,\&\,(Ey)\,\{\ldots\}$		$R(k+1,k+1)$
$+2)^{nd}$ stage		
$R(k+1,k+2)\,\&\,(z)\,[\ldots]$		

The corresponding semantic tableau is (schematically) represented on p. 25. It is convenient, in such a case, to denote the individuals by the numerals '1', '2', '3', …. I wish to lay some stress on the following points.

(i) There is no counter-example with a *finite* universe. For in the first place the universe must not be empty. So we take some individual and give it the name '1'; then we have $\overline{R(1,1)}$ and, furthermore, there must be some individual which fulfils the conditions: $R(1,y)$ and $(z)[R(z,1) \rightarrow R(z,y)]$; as 1 does not fulfil the first condition, the individual we need now must be different from 1. If it is given the name '2', then we clearly have $R(1,2)$ and $(z)[R(z,1) \rightarrow R(z,2)]$. In addition, we have $\overline{R(2,2)}$, and there must be some individual which fulfils the conditions $R(2,y)$ and $(z)[R(z,2) \rightarrow R(z,y)]$; we find that this individual must be different from 1 and 2 and we give it the name '3'. *Etc.*

(ij) In accordance with the discussion under (i), the construction of ou tableau turns out to be fatally determined by the data of the problem. On account of rule (vi b) in section 6, the tableau has to be split up again and again, but most of the subtableaux thus obtained are quickly closed whereas those which are not yield exactly the same truth values for the atomi formulas $R(1,1)$, $R(1,2)$, $R(1,3)$, …, $R(2,1)$, $R(2,2)$, …, $R(3,1)$, …. It is easy to see that the tableau indeed provides us with a counter-example which can be simply described as follows: the universe consists of all natura numbers 1, 2, 3, … and for the binary predicate **R** we take the relatio *smaller than* between natural numbers.

(iij) One might suspect that the emergence of infinite tableaux (and hence, of counter-examples involving infinitely many individuals) points to some deficiency in our approach. However, such a view would be mistaken We wish to establish a logical theory which is adapted to such situations as may present themselves in scientific argument. For instance, it ought t provide the framework for a discussion on the (unsolved) problem whethe or not Fermat's Last Theorem is derivable from the axioms of arithmetic The corresponding semantic tableau would look as follows:

Valid	*Invalid*
Axioms of Arithmetic	Fermat's Last Theorem
—	—
—	—

Now if a counter-example is to satisfy the conditions stated in the le column, it must certainly involve infinitely many individuals. So if in som anner we exclude infinite tableaux, we cannot hope ever to establish an propriate framework for dealing with problems of the above kind.

(3) It will be clear that, generally speaking, with respect to the problem nether or not a certain conclusion V LOGICALLY FOLLOWS from given emisses $U_1, U_2, \ldots$ (the general problem whether or not a certain sequent OLDS TRUE is treated in the same manner), we have to anticipate three ssibilities.

(i) The tableau provides us with a suitable *finite* counter-example for oving that V is *not* semantically entailed by $U_1, U_2, \ldots$; in this case, V is ot derivable in F from $U_1, U_2, \ldots$.

(ij) The construction of the tableau breaks down; in this case V is both mantically entailed by, and derivable in F from $U_1, U_2, \ldots$.

(iij) The construction of the tableau involves infinitely many steps; in this se the tableau is not closed, hence V is not derivable in F from $U_1, U_2, \ldots$.

In cases (i) and (ij), the completeness theorem for the system F clearly olds true. It will also hold true in case (iij) if the construction of the tableau ovides us with a suitable counter-example.

In the particular situation which has just been examined, we were indeed le to point out the existence of a suitable counter-example; however, this uation was exceptionally simple: the counter-example corresponded to a miliar mathematical structure, and so it was recognized rather than discovered.

In general we must be prepared to meet with a chaotic succession of bleaux being split up and of subtableaux being closed. We cannot expect at we will always be able to discover some regularity in the process and en to 'read' some counter-example from the tableau.

Now in order to have at least a guide in this labyrinth, we shall represent ır tableau by a certain configuration which is still more concise and therere gives an even clearer insight into the progress of our systematic attempt constructing a suitable counter-example. This configuration consists of oints and line segments, and is called a (binary) *tree*.

For each formula which we insert in our tableau, one point is added to the orresponding tree. The points corresponding to the formulas in one and e same subtableau are stringwise connected, in accordance with the order which the formulas appear in the tableau, the distinction between the two olumns in a subtableau being disregarded. If the tableau (or subtableau) is lit up, then the tree contains a fork; this shows the relations of subordiation which may exist between various tableaux.

By way of illustration, we have constructed in fig. 2 part of the tree which orresponds to the tableau on p. 25. To save space, we have omitted the arts (a), (b), and (d) which are exactly like (c).

It will be clear that the tree can be considered as composed of *branches* which start at the 'top' and stretch downward as far as possible. Such a branch corresponds to a sequence of 'nested' subtableaux, each subtableau in the sequence being subordinate to all preceding ones. This sequence obviously must belong to one of three types.

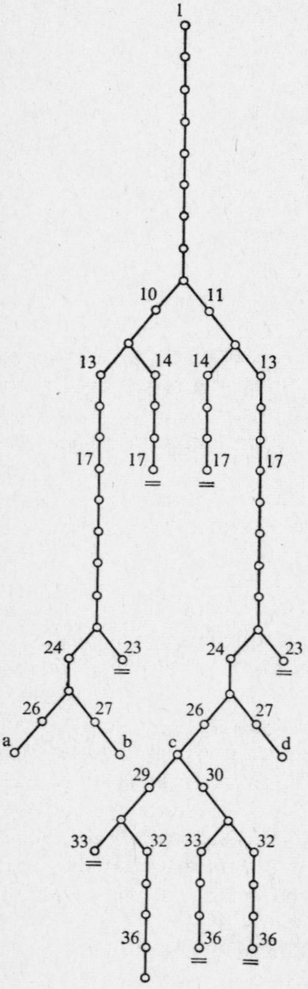

Fig. 2

(i) It may terminate on account of the fact that all possibilities of applying rules (ij)–(vij) in Section 6 have been exhausted.

(ij) It may break off on account of the closure of the last subtableau which it contains.

(iij) It goes on indefinitely; it will be clear that this situation will present itself if and only if the corresponding branch in the tree goes on indefinitely.

Type (ij) is of no importance for us as it does not provide us with a suitable counter-example; however, both type (i) and type (iij) yield suitable counter-examples and hence will be discussed. Actually, we shall only deal with type (iij), as it is very easy to adapt the argument to type (i).

Let us suppose that U_1, U_2, and ... have been the initial formulas in the tableau and that they contain a property letter 'A' and a relation letter 'R'; let the universe consist of the numbers 1, 2, 3, ... We consider the atomic formulas $A(j)$ and $R(j,k)$, $j, k = 1, 2, 3, \ldots$, which appear in the tableau and subtableaux belonging to our sequence. As none of these tableaux is closed, no formula can appear both left and right.

In order to obtain a counter-example we now must select predicates for the letters 'A' and 'R' (cf. Section 1). We take the property **A** and the relation **R** which can be defined as follows.

(A) We say that the natural number k has the *property* **A**, if and only if he atomic formula $A(k)$ appears in a left column (or subcolumn) belonging o the above sequence.

(R) We say that the natural number j is in the *relation* **R** to the natural umber k, if and only if the atomic formula $R(j,k)$ appears in a left column or subcolumn) belonging to the above sequence.

Finally, we have to show that the counter-example thus obtained is a uitable counter-example for proving that the formula V is *not* semantically ntailed by U_1 and U_2. This means that we have to point out that, for the bove property **A** and relation **R** and with respect to the universe which onsists of all natural numbers, the formulas U_1 and U_2 are valid, whereas e formula V is invalid. We do so by showing that *all* formulas in a left olumn (of our sequence) are valid and that *all* formulas in a right column re invalid; let us observe that, for all atomic formulas, this assertion is istified on account of definitions (A) and (R). We now consider non-atomic rmulas; it will be sufficient to discuss a few particular cases, in connection ith the rules for the construction of semantic tableaux.

(iij) Suppose that $(x)X(x)$ appears in a left column, but is invalid; if $)X(x)$ is invalid then, for some natural number k, $X(k)$ must be invalid; n the other hand, all formulas $X(k)$ appear in a left column. It follows that me formula $X(k)$ appears in a left column, but is invalid.

(vi a) Suppose that $X \vee Y$ appears in a left column, but is invalid; in all ontinuations of this left column, either X or Y must appear; on the other and, if $X \vee Y$ is invalid, then both X and Y must be invalid. It follows at either X or Y appears in a left column, but is invalid.

On account of this discussion, any discrepancy between the tableau and e counter-example is transferred to ever shorter formulas, and finally to e atomic formulas. But we know already that, for atomic formulas, there n be no such discrepancy. So we have proved our assertion; it follows that e have indeed obtained a suitable counter-example (in fact, the above scussion merely shows that the rules (i)–(vij) in Section 6 were suitably osen).

So, even if the construction of a semantic tableau involves infinitely many eps, it will nevertheless enable us to point out a suitable counter-example, ovided the corresponding tree contains a branch which goes on indefi- tely. And hence the proof of the completeness theorem for our System F n be concluded by showing that every tree which contains infinitely many oints has a branch of this kind. This can indeed be shown, but it demands digression on the general theory of trees.

8. *A Theorem on Trees.* The configurations, which are known as '*trees*', n be generally characterized as follows. There is one point 1, called

'*origin*', to which we assign the *rank* 1. Furthermore, there are a finit number of points of rank 2, a finite number of points of rank 3, ..., a finit number of points of rank k, and so on; we neither exclude nor require tha starting with a certain rank k, the number of points is 0; it is convenient t introduce, in addition, a *zero-tree*, which does not contain any point.

A point of rank k can be an *endpoint*; otherwise, it is *connected* with a least one point of rank $k+1$. A point of rank $k+1$ is connected with *exactl* one point of rank k. Two points whose ranks are not consecutive can neve be connected.

A sequence of points of ranks 1, 2, 3, ..., such that any two points c consecutive ranks are connected and which, without violating this cor

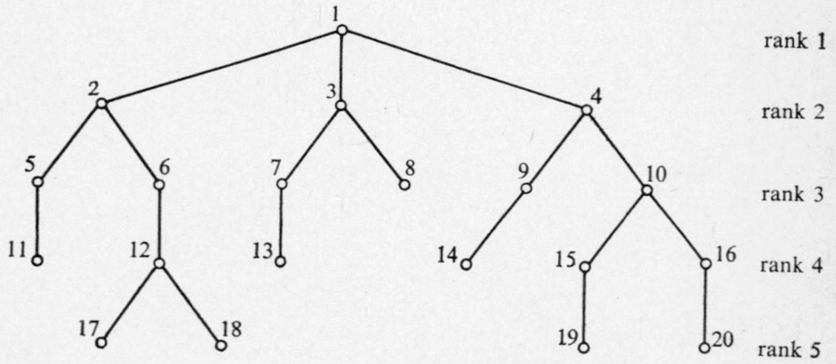

Fig. 3

dition, cannot be extended, is called a *branch* of the tree; such a branc must start at the origin, and it must either break off at an endpoint or go o indefinitely.

Each point P of a tree B clearly is the origin of a *subtree* $B^{(P)}$ of B. $B^{(P)}$ contains only finitely many points, then P is called a *point of the fir kind*; otherwise, P is said to be a *point of the second kind*.

Let B^* be the configuration which remains if, from a tree B, we cancel a points of the first kind.

(1) Let P and Q be connected points of ranks k and $k+1$ in a tree B then we have: (i) if P is of the first kind, then Q is also of the first kind; an (ij) if Q is of the second kind, then P is also of the second kind.

(2) Let P be of rank k in a tree B; then we have: (i) if P is of the secon kind, then it is connected with at least one point Q of rank $k+1$ and of th second kind; and (ij) if all points Q of rank $k+1$ and connected with P a of the first kind, then P is also of the first kind.

(3) The configuration B^* is also a tree.

PROOF. By theorem (1), *sub* (ij), any point Q of rank $k+1$ in B^* is con-ected with at least one point P of rank k in B^*.

(4) No point in B^* can be an endpoint. This follows from (2) *sub* (i).

(5) For a tree B, there are only two possibilities, namely: (i) B contains nly finitely many points, and every branch in B breaks off; and (ij) B ontains infinitely many points, and there is in B at least one branch which oes on indefinitely.

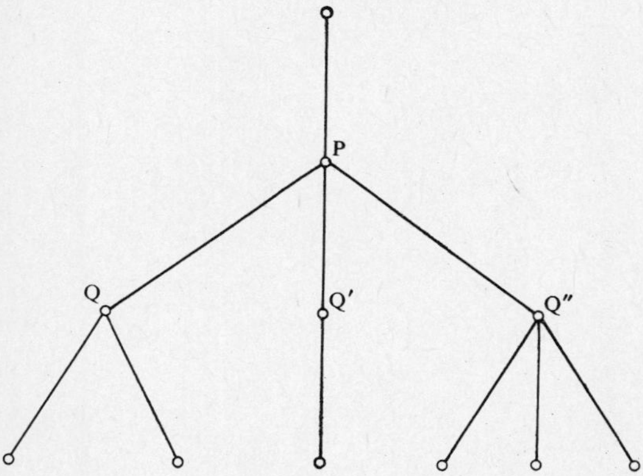

Fig. 4

PROOF. Let us consider the tree B^*. Now there are only two possibilities, amely: (i) B^* is the zero-tree, and (ij) B^* contains at least one point.

d (i) It follows that the origin 1 of B is a point of the first kind so $B^{(1)} = B$ ontains only finitely many points, hence every branch in B breaks off.

d (ij) If B^* contains a point then, by theorem (4), it must contain a branch hich goes on indefinitely. Of course, this branch must also appear in B, d hence B must contain infinitely many points.

It will be clear that theorem (5) fills the last gap in our proof of the mpleteness theorem for the System F.

Notwithstanding the elementary character of the above proof of the '*tree eorem*', it cannot be accepted from an intuitionistic point of view, as it mands an appeal to the principle of the excluded third. It is true that an tuitionistically valid proof can be obtained from Brouwer's proofs of the -called *Fundamental Theorem on Finitary Spreads*, but this requires a ecial interpretation of the theorem by which its application is severely stricted. In fact, my attention was drawn to the tree theorem and to the problems connected with its proof by a discussion with L. Henkin, A. Heyting, and J. J. de Iongh during a seminar on intuitionistic mathematics held in November 1954; this discussion was followed by a conversatio with P. Erdös and L. Henkin on related subjects.

Besides mathematics and botany, many other sciences deal from time t time with trees. In traditional logic, we find the *arbor Porphyriana*, which i also known as the '*Ramean tree*'. According to Jevons,[10] 'Jeremy Benthar speaks truly of "the matchless beauty of the Ramean tree".' In recer philosophical literature, the configuration appears, for instance, in cor nection with the interpretation of Plato's theory of ideal numbers, a defended by J. Stenzel and O. Becker, and in discussions on the intuitionisti theory of continuum.[11]

9. *Range of Application.* In order to indicate the domain of applicatic of the above methods, I shall use the following (summary) division of th domain of modern logic. Let us first make a 'horizontal' distinction betwee (I) classical (or two-valued) logic, (II) intuitionistic and modal logic, ar (III) the systems of many-valued logic. Within each of these domains, the is a 'vertical' gradation into (a) the sentential calculus, (b) elementary logi and (c) logic of higher order. Classical logic is the system which is current applied in scientific argument, and hence it has been studied rath thoroughly. It displays the full hierarchy (a)–(c). The systems under (II) a also sometimes applied, on account of certain objections to classical logi The systems under (III) are experimental constructions, they have hard ever been applied. For the systems under (II) and (III) we have as yet on (a) the sentential calculus, and (b) elementary logic.

The above treatment covered only the classical sentential calculus ar classical elementary logic. It can, however, be extended to the classical log of higher order, and so virtually covers the entire domain of classical logic.[11a] Of course, it remains to be seen, whether a similar treatment is possible f the systems under (II) and (III).

10. *Logical Machines.* The strongly mechanical character of the pr cedures which we have described suggests the possibility of constructing '*logical machine*', which would automatically solve any logical problems the types which have been discussed. The outer part of such a machi would include: a *dial* for 'feeding' the initial formulas to the machine, a

10 W. Stanley Jevons, *The Principles of Science* (reprinted, London, 1920), p. 70

11 W. van der Wielen, *De ideegetallen van Plato* (Amsterdam, 1941); H. We *Philosophy of Mathematics and Natural Science* (Princeton, 1949), p. 53.

11a [This statement of Beth's was somewhat premature; see p. 3 of Introduct above. Ed.]

red, a *yellow*, and a *green light*, one of which would start burning after the ıachine had turned during a certain time.

(a) The red light is meant to announce that the semantic tableau for the roblem is closed; then the machine produces a strip of paper showing the orresponding formal derivation;

(b) The yellow light is meant to announce that the construction of the ıbleau is terminated under rule (x); then the machine produces a strip of aper showing a suitable finite counter-example;

(c) The green light is meant to announce that the construction of the ıbleau involves infinitely many steps; then the machine produces a strip of aper showing some point of B^*.

The inner parts of the machine would have to perform the following ısks.

(A) To construct the semantic tableau with the initial formulas 'fed' to ıe machine;

(B) To construct the corresponding tree B;

(C) To construct the tree B^*.

Unfortunately, it is impossible to construct a machine which performs ıe task under (C). For it will be clear that it would have to construct B^* ›intwise, according as the construction of the tableau and of B progresses. ow let B^k be the part of B which contains the points of rank 1, 2, ..., k, ıd let $B^*_{(k)}$ be the part of B^* which is known as soon as B^k is known.

Then if, for some other tree C, we have $C^k = B^k$, it follows that $B^*_{(k)} = C^*_{(k)}$. ut it is easy to see that C can be chosen in such a manner that it does not ›ntain any points of rank $k+1$. Then C^* is the zero-tree, and so are both $^*_{(k)}$ and $B^*_{(k)}$. Hence it is impossible to construct B^* on the basis of a point-ise construction of B.

It follows that it is impossible to construct a logical machine of the above nd; this conclusion agrees with a famous (and completely general) eorem which was published in 1936 by A. Church. However, it is possible construct a logical machine with a red and a yellow light, and such a achine would already be extremely helpful. For instance, it would enable ; to check any given derivation, and to simplify it, if possible.

The above criticism of the conception of a logical machine also affects, to certain extent, the proof which has been given of the tree theorem. This oof presupposes a '*platonistic*' notion of infinity; for if we accept B^* as a ell-defined entity, this means that we assume the possibility of visualizing not only pointwise but also as a totality. Hence intuitionism, which rejects ch platonistic notions, cannot accept the above proof. On the other hand, is not possible to prove the completeness theorem on the basis of weaker sumptions for, conversely, the completeness theorem implies the tree theorem. It is possible, however, to give an intuitionistically valid proof of a weaker version of the completeness theorem (J. Herbrand, 1930).[12]

The logical machine with a red and a yellow light displays a striking analogy to the reasoning power of the human intellect and, in fact, it is able to perform a considerable number of its operations and even to perform them with a higher degree of efficiency and of accuracy. However, the human intellect is equipped with additional operations which go beyond the power of such a logical machine. For instance, it is not difficult to construct the tree B^*, if B is the tree of fig. 2; so it seems that our intellect is equipped with a kind of green light. Nevertheless, we may not assume that our intellect operates this green light in a systematic and hence perfectly reliable manner for this would imply that the human intellect incorporated, so to speak, logical machine with a red, a yellow, and a green light, and we know that such a machine cannot exist.

11. *Historical Remarks—Traditional Logic and Symbolic Logic.* The above results may have given some impression of the progress made by symbolic logic since Land's communication on this subject.[13] At the same time, the gap between symbolic and traditional logic has considerably narrowed;[14] this is not due to the influence of regressive tendencies in contemporary logic, but rather to the development of the systems of natural deduction, which have enabled symbolic logic to incorporate certain notions which are already found in traditional logic but which remained foreign to symbolic logic in its earlier stages. In fact, none of the procedures which have been applied in the present communication: the introduction of special symbols, and the construction of counter-examples, of semantic tableaux, of formal derivations, and of trees, were entirely unknown

12 It may be helpful to give a brief statement concerning the effect of the above criticism of our proof of the completeness theorem (which, on account of the impossibility of proving this theorem on the basis of weaker assumptions, applies to any proof of it).

(i) Our criticism does not affect the usual proofs of the converse theorem, according to which formal derivability implies semantic entailment (cf. Section 2; we did not return to this theorem, as it is obvious in the context of our treatment).

(ij) If we reject the proofs of the completeness theorem, this does not compel us to consider this theorem as false.

(iij) Suppose the completeness theorem to be true, but not provable (that is, not formally derivable from such assumptions as were tacitly made in the above exposition); then, by the completeness theorem itself, there must be a suitable counter-example for showing that it is not provable. That is, there is an interpretation with respect to which the completeness theorem is invalid (this interpretation must be different from the 'normal' one; for the completeness theorem was supposed to be true, that is, valid with respect to its 'normal' interpretation).

13 J. P. N. Land, 'On a Supposed Improvement in Formal Logic', *Versl. & Med. R. 5*, 1876.

14 J. T. Clark, *Conventional Logic and Modern Logic* (Woodstock, Md., 1952).

raditional logic; but nowadays such devices are more systematically applied ınd more thoroughly analysed. This factor, besides other ones (in particular, he interaction of logical research with investigations into the foundations of mathematics), explains the superiority of modern symbolic logic over raditional logic which appears, for instance, from the following concrete acts:

(i) Symbolic logic provides a more detailed analysis for forms of argunent.

(ij) It shows that certain problems which were vividly discussed in traitional logic derive from inaccurate notions and that, if these notions are orrected, these problems simply vanish.

(iij) Certain other problems, rightly asked by traditional logic but not olved in a satisfactory manner, are solved by modern logic.

(iv) Symbolic logic has discovered certain fundamental problems which raditional logic overlooked on account of its failure to make a clear disinction between the two notions of LOGICAL CONSEQUENCE as discussed in ection 1.

d (i) As an example I mention the above analysis of the syllogism in ESTINO. We have dissolved this type of argument into *nine* successive steps. s all formulas involved were dissolved into atomic formulas, it is pretty lear that a more detailed analysis is impossible. Now none of the steps (or ules of inference) which are found in our analysis were unknown in traitional logic; furthermore the general idea which underlies our analysis ully agrees with the principle of Aristotle's method of proof by ἔκθεσις.[15] Nevertheless, traditional logic never succeeded in giving a complete nalysis of the different modi of the syllogism.

d (ij) A problem which is frequently discussed in traditional logic is conerned with the possibility of reducing the so-called hypothetical syllogism o the categorical syllogism.[16] Now this problem cannot arise if the situation correctly understood. Both the categorical and the hypothetical syllogism an be dissolved into (and hence reduced to) a certain number of rules of nference of a more elementary character. The possibility of such an analysis vas already clearly seen by Aristotle himself. The rules of inference which lay a role in the analysis of the hypothetical syllogism belong exclusively to he domain of the sentential calculus;[17] an analysis of the categorical

[15] Aristotle, *Prior Analytics* A 6, 28^b 17, applies this method in the proof of the odus BOCARDO, which is closely related to the modus FESTINO.

[16] Th. A. van den Berg, *Het hypothetisch denken* (Nijmegen, 1954).

[17] The first traces of this logical discipline are found in the later works of Aristotle, ut its development is due to his disciples Eudemus and Theophrastus and, in parcular, to the Stoics. Cf. Bocheński, l.c., and, for more details: Benson Mates, *Stoic ogic* (Berkeley-Los Angeles, 1953).

syllogism involves, *in addition*, certain rules of inference which belong t
the theory of quantifiers. Hence the hypothetical syllogism is certainly of
more elementary character, but of course it does not follow that it can b
reduced to the categorical syllogism.

ad (iij) Let us consider a situation which frequently arises in mathematica argumentation. Suppose, for instance, that we wish to establish a certai theorem on chordal quadrangles. Then we might start as follows: Let Γ b a given circle, and let *PQRS* be a given quadrangle inscribed in it; the adding these suppositions to the 'given premisses' (namely, the axioms c geometry), we might find that the quadrangle *PQRS* has some peculia property **E**. And then we would argue: the special choice of the circle Γ doe not matter at all; the only point is that, for the quadrangle *PQRS*, such circle Γ can be found. Accordingly, we would state our conclusion a follows: Any quadrangle *PQRS*, for which there is a circle Γ passin through the four vertices *P*, *Q*, *R*, and *S*, has the property **E**.

Aristotle[18] was presumably the first to discuss this type of argument:

> Therefore, it has also been said that it is necessary to suppose som thing false, just as even the geometers suppose to be a foot long wha is not a foot long. But this cannot be the case. For the geometers do nc suppose anything which is false (for in the formal argument such premiss does not occur) ....

The above example gives an illustration of what is meant. In a certai part of the proof it may appear as if the quadrangle *PQRS* ought to b inscribed in the particular circle Γ, and this of course is not correct. So or might suspect that the proof of the theorem involves an appeal to a fal assumption. A similar situation arises in the formal derivation which w discussed in Section 4, after hypothesis 1 has been introduced. In all the cases, however, the (possibly false) assumption, which at a certain mome has been introduced, is eliminated later on; so it does not count as a premis and this is exactly what Aristotle observes. However, if we wish exactly t know what is going on, then we ought to consult the semantic tableau. I the formal derivation of Section 4, we know by premiss (2), that son individual fulfils the condition $S(y)$ & $M(y)$, and we agree to give th individual the name '*a*'. Likewise, if for *PQRS* there is a circle in which it inscribed, then we may agree to give this circle the name 'Γ'; and, i Aristotle's example, if we are given a certain line segment, why should n we take it as a unity of length?

18 Aristotle, *Metaphysics* N 2, 1089^a 21–25; this text is connected with *Metaphysi* M 3, 1078^a 17–20, and, less closely, with *Posterior Analytics* A 4, 73^b 30–3.

Considerable attention has been given to this matter by Locke, Berkeley, Hume, and Kant.[19] The views defended by Locke, Berkeley, and Hume are reasonable enough, even though, naturally, a detailed logical analysis of the situation was beyond their power. According to Kant, the introduction of an individual (Γ, or a) is entailed by the necessity of connecting the formal argument with the construction of such an individual; hence this type of argument involves an appeal to spatial intuition, it is characteristic of mathematics and cannot be applied in other fields. Curiously enough, a similar remark is made by Alexander in connection with Aristotle's method of proof by ἔκθεσις: he asserts that this method involves an appeal to perceptual evidence.[20] Such conceptions, however, are clearly inconsistent with the insight which we easily derive from an inspection of the semantic tableau.

Another problem which has been discussed in traditional logic but which only symbolic logic has been able to treat in an efficient manner is concerned with the analysis of arguments which involve relations. We have seen in Section 7, that the analysis of arguments of this kind leads to considerable complications. Strangely enough, it is still maintained by some authors that it is easy to reduce such arguments to categorical syllogisms.[21]

ad (iv) The two notions of LOGICAL CONSEQUENCE were (implicitly) used long before they were clearly distinguished and correctly defined. In logical theory, people always tended to use the notion of formal derivability. The method of the counter-example and hence the notion of semantic entailment was used in philosophical discussion.[22] The development of symbolic logic resulted, of course, in the first place in a clarification of the notion of formal derivability. But between 1920 and 1930, the distinction between the notions of derivability theory and those of semantics (or satisfiable theory) was gradually adopted and this enabled Gödel[3] to state and solve the completeness

[19] J. Locke, *An Essay concerning Human Understanding*, Book IV, Chapter 7, § 9; G. Berkeley, *A Treatise concerning the Principles of Human Knowledge*, Introduction §§ XII–XVI; cf. F. Bender, *George Berkeley's Philosophy re-examined* (Amsterdam, 1946); D. Hume, *A Treatise of Human Nature*, Vol. I, Book I, Part I, § 7; I. Kant, *Kritik der reinen Vernunft*, A 25, A 320, A 713–14; cf. E. W. Beth, Kants Einteilung der Urteile in analytische und synthetische, *Alg. Ned. Tijdschr. v. Wijsbeg. en Psychol.*, Vol. 46 (1953–4), pp. 253–64.

From the recent literature on the subject, I mention: K. Kroman, *Unsere Naturerkenntnis* (Kopenhagen, 1883); E. Husserl, *Logische Untersuchungen*, 2.Bd, I. Teil (1901), 2 Aufl. (Halle a.d. S. 1913); E. Mach, *Erkenntnis und Irrtum* (1905), 2. Aufl. (Leipzig, 1906); O. Hölder, *Die mathematische Methode* (Berlin, 1924).

[20] Cf. J. Łukasiewicz, l.c.[5], p. 60.

[21] J. Maritain, *Petite logique* (Eléments de philosophie, II: L'ordre des concepts), 11me éd. (Paris, 1933), pp. 197–8.

[22] An early example is found in Alexinus' objections to Zeno's argument in support of his conception of the Universe as a thinking and animate Being; cf. Sextus Empiricus, *Adversus mathematicos* IX, 104–9.

problem for elementary logic. The first to give a satisfactory definition of the notion of semantic entailment was Tarski.[23]. Without the background of symbolic logic, nobody could ever have been aware of the importance of the completeness problem.

12. *Logic and Mathematics.* It is frequently argued that symbolic logic lays undue stress upon the connections between logic and mathematics and, in fact, tends to making logic subservient to mathematics. Let us try to judge the foundedness of such accusations; the above discussion, which has been concerned with a number of vital topics in modern logic, should provide us with a safe basis for our verdict.

I hope that it has been made sufficiently clear that, though modern logic uses numerous devices (symbols, formulas, tableaux, configurations) which strongly recall those which are currently used in mathematics, this ought not to be blamed upon the influence of mathematical conceptions, but rather ascribed to the technical needs of logic which happen to be in many respects quite similar to those of mathematics. Many of these devices were already applied in traditional logic, though usually in a less sophisticated form; to many of them we naturally resort when dealing, in every-day life, with a more or less involved situation. Such devices as are used in logic are never merely copied from those which one finds in mathematics, they are always adapted to the special demands of logical research.

There is, however, one step in the above argument, where an overt appeal was made to the resources of mathematics: it is found in the proof of our completeness theorem, where we applied the tree theorem; as both the statement and the proof of the tree theorem involve the notion of infinity, it seems that its mathematical character can hardly be denied.

It is certainly true that this step creates a curious situation; but its implications are by no means as obvious as one may believe.

(i) The tree theorem itself is open to criticism and so one might certainly contemplate the possibility of replacing it, in the proof of the completeness theorem, by some suitable postulate, which then could have a purely logical character.

(ij) In many cases one can use, instead of the full completeness theorem, one of its weaker versions, the proofs of which do not present the above-mentioned difficulties.

(iij) The more important applications of the completeness theorem are

23 A. Tarski, 'Ueber den Begriff der logischen Folgerung', *Actes du Congrès International de Philosophie Scientifique*, fasc. VII: *Logique* (Paris, 1936), pp. 1–11; as a forerunner, one may mention B. Bolzano, *Wissenschaftslehre* (Sulzbach, 1837), 2.Bd p. 198.

ll concerned with the application of logical methods in research on the oundations of mathematics. It will be better briefly to describe the charac-er of these applications. Let U_1, U_2, ... be the axioms of a certain mathe-natical theory T; it may happen that we do not know a system of entities or which these axioms are valid (such a system of entities is currently enoted as a *model* for the axioms) although we know that the axioms are ormally consistent. Then we consider the semantic tableau.

	Valid	*Invalid*
(1)	U_1	—
(2)	U_2	—
	—	—

As U_1 U_2 ... are formally consistent (with respect to the system F), it ollows that the tableau cannot be closed. Hence it will provide a suitable ounter-example for proving that the sequent:

$$U_1, U_2, \ldots \vdash \emptyset$$

loes *not* hold true; it is easy to see that this counter-example is at the same ime a *model* for the axioms U_1, U_2, ... of the theory T. It will be clear that, n connection with problems of this kind, the question of methodical purity an hardly be raised.

(iv) It is difficult to accept the thesis that the notion of infinity is entirely oreign to pure logic; in fact, many treatises of traditional logic contain nore or less extensive digressions on the subject. And, for instance, in liscussing those semantic tableaux which are not closed, we can hardly be orbidden to make a distinction between those tableaux which terminate inder rule (x), and those which go on indefinitely, and there already the notion of infinity presents itself in a quite natural manner. On the other hand, if we try to assign the tree theorem its place within the field of (pure) nathematics, it turns out to belong to a domain which has an extremely bstract character and is strongly related to pure logic. In fact, the central nd more elementary parts of logic are clearly distinct from the central and nore elementary parts of mathematics, but if we pass on to the more idvanced parts of the two fields it is sometimes very difficult to maintain a lear separation. Very often it depends mainly on the wording of a theorem o which of the two fields it will be counted; and this applies in particular to he tree theorem.

Under these circumstances, it seems wiser not to take the above accusations too tragically. But I wish to emphasize that these accusations are not justified if they are meant to affect the *principles* of the two disciplines involved. If there is a certain degree of mutual penetration of the two fields, it is confined to their more outlying parts. And there it is, as far as I can see, both innocuous and difficult to avoid.

At the end of this communication I wish to extend my thanks to Professors Paul Bernays, I. M. Bocheński, and Leon Henkin, to Dr. Gisbert Hasenjaeger, and to my assistant, Mr. S. J. Doorman, for the interest they have given me during its preparation.

POSTSCRIPT

(*Added July 11, 1955*) Since, at the meeting of the Section of Letters of May 9, 1955, I presented the above communication, a few developments have taken place on which I should like briefly to report.

(1) Professor Quine, to whom I sent a copy of my manuscript, has kindly drawn my attention upon two papers by K. J. J. Hintikka.[24] Hintikka's 1955 paper indeed contains ideas which are very closely related to those underlying my present communication. In particular the Author also stresses the interpretation of all proofs of logical truth as proofs of impossibility of counter-examples (p. 49) and resorts to the introduction of a certain tree (p. 38). However, he does not use semantic tableaux and hence he is compelled to apply instead a certain formal technique which consists, roughly in the extension of any consistent set of (closed) formulas into a complete and consistent 'normal' set; this technique (which was independently developed by G. Hasenjaeger and by L. Henkin) was—as far as I know—first described in my own 1951 paper.[4] Hintikka rightly points out its connections with Herbrand's and Gentzen's results; this point was briefly raised in my own 1953 paper 'On Peano's Method'.[4] I gave a systematic treatment of the whole subject in my lectures at the Sorbonne in Spring, 1954.[25]—It needs hardly saying that Hintikka's two papers also contain much material which is found neither in my previous publications nor in my present communication.

On the basis of his results, the author states (p. 49) the redundancy of the so-called negationless logics; this point has been previously treated by P. C

[24] K. Jaakko J. Hintikka, *Distributive Normal Forms in the Calculus of Predicate* (Acta Philosophica Fennica, Fasc. VI, Helsinki, 1953); *Two Papers on Symboli Logic* (ibid., Fasc. VIII, Helsinki, 1955; in particular the first paper, 'Form an Content in Quantification Theory').

[25] E. W. Beth, *L'existence en mathématiques* (Paris, 1955).

Gilmore.[26] From an intuitionistic point of view, however, neither Hintikka's nor my own treatment of semantic entailment is conclusive. In the near future I hope to publish the intuitionistic counter-part of my present communication.

(2) In my 'Remarks on Natural Deduction',[4] I stated the conjecture that semantic tableaux might provide us with a decision procedure for intuitionistic validity of classically valid formulas. I am indebted to Professor S. C. Kleene for the following argument which shows that a relative decision procedure of this kind cannot exist.

Suppose we had such a formula, and let X be *any* formula. Let Y be some formula known as classically valid and intuitionistically invalid. The formula $X \vee Y$ will then at any rate be classically valid. We apply to it our supposed relative decision procedure. Now if $X \vee Y$ turns out to be intuitionistically invalid, then clearly X must also be intuitionistically invalid (for intuitionistically X entails $X \vee Y$). On the other hand, if $X \vee Y$ turns out to be intuitionistically valid, then X must be intuitionistically valid (for, by a well-known result by Gödel, $X \vee Y$ cannot be intuitionistically valid, unless either X or Y is intuitionistically valid). So we would have an absolute decision procedure for intuitionistic validity. But by the analogue of Church's theorem for intuitionistic logic such an absolute decision procedure cannot exist.

(3) Finally, I wish to mention the publication of a French translation, with useful comments, of Gentzen's 'Untersuchungen',[27]

[26] P. C. Gilmore, 'The Effect of Griss' Criticism of the Intuitionistic Logic on Deductive Theories Formalized within the Intuitionistic Logic' (Thesis, University of Amsterdam; Amsterdam, 1953).

[27] G. Gentzen, *Recherches sur la Déduction logique*, traduction et commentaire par R. Feys et J. Ladrière (Paris, 1955).

II

THE COMPLETENESS OF THE FIRST-ORDER FUNCTIONAL CALCULUS

LEON HENKIN[1]

ALTHOUGH several proofs have been published showing the completeness o the propositional calculus (cf. Quine[2]), for the first-order functional cal culus only the original completeness proof of Gödel[3] and a variant due t Hilbert and Bernays have appeared. Aside from novelty and the fact that i requires less formal development of the system from the axioms, the ne method of proof which is the subject of this paper possesses two advantages In the first place an important property of formal systems which is associ ated with completeness can now be generalized to systems containing non-denumerable infinity of primitive symbols. While this is not of especia interest when formal systems are considered as *logics*—i.e., as means fo analyzing the structure of languages—it leads to interesting applications i the field of abstract algebra. In the second place the proof suggests a ne approach to the problem of completeness for functional calculi of highe order. Both of these matters will be taken up in future papers.

The system with which we shall deal here will contain as primitive symbo

$$(\quad) \quad \supset \quad f \quad ,$$

and certain sets of symbols as follows:

(i) *propositional symbols* (some of which may be classed as *variable* others as *constants*), and among which the symbol '*f*' above is to be i cluded as a constant;

(ii) for each number $n = 1, 2, \ldots$ a set of *functional symbols of degree* (which again may be separated into *variables* and *constants*); and

From the *Journal of Symbolic Logic*, Vol. 14 (1949), pp. 159–66. Copyright © 194 Reprinted by permission of the publishers, the American Mathematical Society, ar the author.

[1] This paper contains results of research undertaken while the author was a Nation Research Council predoctoral fellow. The material is included in 'The Completene of Formal Systems', a thesis presented to the faculty of Princeton University candidacy for the degree of Doctor of Philosophy.

[2] W. V. Quine, 'Completeness of the Propositional Calculus', *Journal of Symbo Logic*, Vol. 3 (1938), pp. 37–40.

[3] Kurt Gödel, 'Die Vollständigkeit der Axiome des logischen Funktionenkalkül *Monatshefte für Mathematik und Physik*, Vol. 37 (1930), pp. 349–60.

(iii) *individual symbols* among which *variables* must be distinguished from *onstants*. The set of variables must be infinite.

Elementary well-formed formulas are the propositional symbols and all ormulas of the form $G(x_1, \ldots, x_n)$ where G is a functional symbol of degree and each x_i is an individual symbol.

Well-formed formulas (wffs) consist of the elementary well-formed fornulas together with all formulas built up from them by repeated application f the following methods:

(i) If A and B are wffs so is $(A \supset B)$.

(ii) If A is a wff and x an individual variable then $(x)A$ is a wff. Method i) for forming wffs is called *quantification with respect to the variable x*. Any ccurrence of the variable x in the formula $(x)A$ is called *bound*. Any ccurrence of a symbol which is not a bound occurrence of an individual ariable according to this rule is called *free*.

In addition to formal manipulation of the formulas of this system we shall e concerned with their *meaning* according to the following interpretation. he propositional constants are to denote one of the truth values, T or F, he symbol 'f' denoting F, and the propositional variables are to have the et of these truth values as their range. Let an arbitrary set, I, be specified s a domain of individuals, and let each individual constant denote a paricular element of this domain while the individual variables have I as their ange. The functional constants (variables) of degree n are to denote (range ver) subsets of the set of all ordered n-tuples of I. $G(x_1, \ldots, x_n)$ is to have he value T or F according as the n-tuple $(x_1, \ldots, x_n)$ of individuals is or is ot in the set G; $(A \supset B)$ is to have the value F if A is T and B is F, otherwise ; and $(x)A$ is to have the value T just in case A has the value T for every lement x in I.[4]

If A is a wff, I a domain, and if there is some assignment of denotations to he constants of A and of values of the appropriate kind to the variables vith free occurrences in A, such that for this assignment A takes on the value according to the above interpretation, we say that A is *satisfiable with espect to I*. If *every* such assignment yields the value T for A we say that A *valid with respect to I*. A is *valid* if it is valid with respect to every domain. Ve shall give a set of axioms and formal rules of inference adequate to ermit formal proofs of every valid formula.

Before giving the axioms, however, we describe certain rules of

[4] A more precise, syntactical account of these ideas can be formulated along the nes of Tarski (Alfred Tarski, 'Der Wahrheitsbegriff in den formalisierten Sprachen', *tudia Philosophica*, Vol. 1 (1936), pp. 261–405). But this semantical version will uffice for our purposes.

abbreviation which we use to simplify the appearance of wffs and formula schemata. If A is any wff and x any individual variable we write

$$\sim A \quad \text{for} \quad (A \supset f),$$

$$(\exists x) A \quad \text{for} \quad \sim (x) \sim A.$$

From the rules of interpretation it is seen that $\sim A$ has the value T or F according as A has the value F or T, while $(\exists x) A$ denotes T just in case there is some individual x in I for which A has the value T.

Furthermore we may omit outermost parentheses, replace a left parenthesis by a dot omitting its mate at the same time if its mate comes at the end of the formula (except possibly for other right parentheses), and put a sequence of wffs separated by occurrences of '$\supset$' when association to the left is intended. For example,

$$A \supset B \supset . \, C \supset D \supset E \quad \text{for} \quad ((A \supset B) \supset ((C \supset D) \supset E)),$$

where A, B, C, D, E may be wffs or abbreviations of wffs.

If A, B, C are any wffs, the following are called *axioms*:

1. $C \supset . \, B \supset C$
2. $A \supset B \supset . \, A \supset (B \supset C) \supset . \, A \supset C$
3. $A \supset f \supset f \supset A$
4. $(x)(A \supset B) \supset . \, A \supset (x) B$, where x is any individual variable with no free occurrence in A.
5. $(x) A \supset B$, where x is any individual variable, y any individual symbol, and B is obtained by substituting y for each free occurrence of x in A, provided that no free occurrence of x in A is in a well-formed part of A of the form $(y) C$.

There are two formal rules of inference:

I (*Modus Ponens*). To infer B from any pair of formulas A, $A \supset B$.
II (*Generalization*). To infer $(x) A$ from A, where x is any individual variable.

A finite sequence of wffs is called a *formal proof from assumptions* Γ, where Γ is a set of wffs, if every formula of the sequence is either an axiom, an element of Γ, or else arises from one or two previous formulas of the sequence by *modus ponens* or generalization, except that no variable with a free occurrence in some formula of Γ may be generalized upon. If A is the last formula of such a sequence we write $\Gamma \vdash A$. Instead of $\{\Gamma, A\} \vdash B$ ($\{\Gamma, A\}$ denoting the set formed from Γ by adjoining the wff A), we shall write $\Gamma, A \vdash B$. If Γ is the empty set we call the sequence simply a *formal proof* and write $\vdash A$. In this case A is called a *formal theorem*. Our object is to show that every valid formula is a formal theorem, and hence that our system of axioms and rules is *complete*.

The following theorems about the first-order functional calculus are all her well-known and contained in standard works, or else very simply rivable from such results. We shall use them without proof here, referring e reader to Church[5] for a fuller account.

III (*The Deduction Theorem*). If $\Gamma, A \vdash B$ then $\Gamma \vdash A \supset B$ (for any wffs A, and any set Γ of wffs).

6. $\vdash B \supset f \supset . B \supset C$
7. $\vdash B \supset . C \supset f \supset . B \supset C \supset f$
8. $\vdash (x)(A \supset f) \supset . (\exists x) A \supset f$
9. $\vdash (x) B \supset f \supset . (\exists x)(B \supset f)$

IV. If Γ is a set of wffs no one of which contains a free occurrence of the dividual symbol u, if A is a wff and B is obtained from it by replacing ch free occurrence of u by the individual symbol x (none of these occurnces of x being bound in B), then if $\Gamma \vdash A$, also $\Gamma \vdash B$.

This completes our description of the formal system; or, more accurately, a class of formal systems, a certain degree of arbitrariness having been t with respect to the nature and number of primitive symbols.

Let S_0 be a particular system determined by some definite choice of imitive symbols. A set Λ of wffs of S_0 will be called *inconsistent* if $\Lambda \vdash f$, herwise *consistent*. A set Λ of wffs of S_0 will be said to be *simultaneously tisfiable* in some domain I of individuals if there is some assignment of notations (values) of the appropriate type to the constants (variables) th free occurrences in formulas of Λ, for which each of these formulas s the value T under the interpretation previously described.

THEOREM. *If Λ is a set of formulas of S_0 in which no member has any currence of a free individual variable, and if Λ is consistent, then Λ is. nultaneously satisfiable in a domain of individuals having the same cardinal mber as the set of primitive symbols of S_0.*

We shall carry out the proof for the case where S_0 has only a denumerable finity of symbols, and indicate afterward the simple modifications needed the general case.

Let u_{ij} $(i, j = 1, 2, 3, \ldots)$ be symbols not occurring among the symbols of . For each i $(i = 1, 2, 3, \ldots)$ let S_i be the first-order functional calculus 1ose primitive symbols are obtained from those of S_{i-1} by adding the mbols u_{ij} $(j = 1, 2, 3, \ldots)$ as individual constants. Let S_ω be the system 1ose symbols are those appearing in any one of the systems S_i. It is easy see that the wffs of S_ω are denumerable, and we shall suppose that some

[5] Alonzo Church, *Introduction to Mathematical Logic, Part I,* Annals of Mathetics Studies, Princeton University Press, 1944.

particular enumeration is fixed on so that we may speak of the first, second ..., nth, ... formula of S_ω in the standard ordering.

We can use this ordering to construct in S_0 a maximal consistent set o cwffs, Γ_0, which contains the given set Λ. (We use 'cwff' to mean *closed* wff *a* wff *which contains no free occurrence of any individual variable.*) Γ_0 i maximal consistent in the sense that if A is any cwff of S_0 which is not in Γ then Γ_0, $A \vdash f$; but not $\Gamma_0 \vdash f$.

To construct Γ_0 let Γ_{00} be Λ and let B_1 be the first (in the standar ordering) cwff A of S_0 such that $\{\Gamma_{00}, A\}$ is consistent. Form Γ_{01} by addin B_1 to Γ_{00}. Continue this process as follows. Assuming that Γ_{0i} and B_i hav been found, let B_{i+1} be the first cwff A (of S_0) after B_i, such that $\{\Gamma_{0i}, A\}$ i consistent; then form Γ_{0i+1} by adding B_{i+1} to Γ_{0i}. Finally let Γ_0 be com posed of those formulas appearing in any Γ_{0i} $(i = 0, 1, \ldots)$. Clearly Γ *contains* Λ. Γ_0 is *consistent*, for if $\Gamma_0 \vdash f$ then the formal proof of f fro assumptions Γ_0 would be a formal proof of f from some finite subset of Γ as assumptions, and hence for some i $(i = 1, 0, \ldots)$ $\Gamma_{0i} \vdash f$ contrary t construction of the sets of Γ_{0i}. Finally, Γ_0 is *maximal* consistent becaus if A is a cwff of S_0 such that $\{\Gamma_0, A\}$ is consistent then surely $\{\Gamma_{0i}, A\}$ i consistent for each i; hence A will appear in some Γ_{0i} and so in Γ_0.

Having obtained Γ_0 we proceed to the system S_1 and form a set Γ_1 of i cwffs as follows. Select the first (in the standard ordering) cwff of Γ_0 whic has the form $(\exists x)A$ (unabbreviated: $((x)(A \supset f) \supset f)$), and let A' be th result of substituting the symbol u_{11} of S_1 for all free occurrences of th variable x in the wff A. The set $\{\Gamma_0, A'\}$ must be a consistent set of cwf of S_1. For suppose that Γ_0, $A' \vdash f$. Then by III (the Deduction Theorem $\Gamma_0 \vdash A' \supset f$; hence by IV, $\Gamma_0 \vdash A \supset f$; by II, $\Gamma_0 \vdash (x)(A \supset f)$; and so by and I, $\Gamma_0 \vdash (\exists x)A \supset f$. But by assumption $\Gamma_0 \vdash (\exists x)A$. Hence mod ponens gives $\Gamma_0 \vdash f$ contrary to the construction of Γ_0 as a consistent set

We proceed in turn to each cwff of Γ_0 having the form $(\exists x)A$, and for t j^{th} of these we add to Γ_0 the cwff A' of S_1 obtained by substituting th constant u_{1j} for each free occurrence of the variable x in the wff A. Each these adjunctions leaves us with a consistent set of cwffs of S_1 by the arg ment above. Finally, after all such formulas A' have been added, we enlar the resulting set of formulas to a maximal consistent set of cwffs of S_1 i the same way that Γ_0 was obtained from Λ in S_0. This set of cwffs we call Γ

After the set Γ_i has been formed in the system S_i we construct Γ_{i+1} i S_{i+1} by the same method used in getting Γ_1 from Γ_0 but using the constan u_{i+1j} $(j = 1, 2, 3, \ldots)$ in place of u_{1j}. Finally we let Γ_ω be the set of cwffs S_ω consisting of all those formulas which are in any Γ_i. It is easy to s that Γ_ω possesses the following properties:

(i) Γ_ω is a maximal consistent set of cwffs of S_ω.

(ii) If a formula of the form $(\exists x)A$ is in Γ_ω then Γ_ω also contains a rmula A' obtained from the wff A by substituting some constant u_{ij} for ch free occurrence of the variable x.

Our entire construction has been for the purpose of obtaining a set of rmulas with these two properties; they are the only properties we shall e now in showing that the elements of Γ_ω are simultaneously satisfiable in denumerable domain of individuals.

In fact we take as our domain I simply the set of individual constants of $_\omega$, and we assign to each such constant (considered as a symbol in an terpreted system) itself (considered as an individual) as denotation. It mains to assign values in the form of truth-values to propositional mbols, and sets of ordered n-tuples of individuals to functional symbols degree n, in such a way as to lead to a value T for each cwff of Γ_ω.

Every propositional symbol, A, of $\boldsymbol{S}_0$ is a cwff of $\boldsymbol{S}_\omega$; we assign to it the lue T or F according as $\Gamma_\omega \vdash A$ or not. Let G be any functional symbol degree n. We assign to it the class of those n-tuples $\langle a_1, \ldots, a_n \rangle$ of indidual constants such that $\Gamma_\omega \vdash G(a_1, \ldots, a_n)$.

This assignment determines a unique truth-value for each cwff of $\boldsymbol{S}_\omega$ der the fundamental interpretation prescribed for quantification and '$\supset$'. /e may note that the symbol 'f' is assigned F in agreement with that erpretation since Γ_ω is consistent.) We now go on to show the

LEMMA. *For each* cwff *A of $\boldsymbol{S}_\omega$ the associated value is* T *or* F *according as* $_\omega \vdash A$ *or not.*

The proof is by induction on the length of A. We may notice, first, that ve do not have $\Gamma_\omega \vdash A$ for some cwff A of $\boldsymbol{S}_\omega$ then we do have $\Gamma_\omega \vdash A \supset f$. r by property (i) of Γ_ω we would have $\Gamma_\omega, A \vdash f$ and so $\Gamma_\omega \vdash A \supset f$ by

.

In case A is an elementary cwff the lemma is clearly true from the nature of assignment.

Suppose A is $B \supset C$. If C has the value T, by induction hypothesis $\Gamma_\omega \vdash C$; n $\Gamma_\omega \vdash B \supset C$ by 1 and I. This agrees with the lemma since $B \supset C$ has value T in this case. Similarly, if B has the value F we do not have $\Gamma_\omega \vdash B$ induction hypothesis. Hence $\Gamma_\omega \vdash B \supset f$, and $\Gamma_\omega \vdash B \supset C$ by 6 and I. ain we have agreement with the lemma since $B \supset C$ has the value T in s case also. Finally if B and C have the values T and F respectively, so t (induction hypothesis) $\Gamma_\omega \vdash B$ while $\Gamma_\omega \vdash C \supset f$, we must show that $\vdash B \supset C$ does not hold (since $B \supset C$ has the value F in this case). But 7 and two applications of I we conclude that $\Gamma_\omega \vdash B \supset C \supset f$. Now we that if $\Gamma_\omega \vdash B \supset C$ then $\Gamma_\omega \vdash f$ by I, contrary to the fact that Γ_ω is nsistent (property i).

Suppose A is $(x)B$. If $\Gamma_\omega \vdash (x)B$ then (by 5 and I) $\Gamma_\omega \vdash B'$, where B' i obtained by replacing all free occurrences of x in B by some (arbitrary individual constant. That is (induction hypothesis), B has the value T fo every individual x of I; therefore A has the value T and the lemma i established in this case. If, on the other hand, we do not have $\Gamma_\omega \vdash (x)I$ then $\Gamma_\omega \vdash (x)B \supset f$ whence (by 9, I) $\Gamma_\omega \vdash (\exists x)(B \supset f)$. Hence, by propert ii of Γ_ω, for some individual constant u_{ij} we have $\Gamma_\omega \vdash B' \supset f$, where E is obtained from B by replacing each free occurrence of x by u_{ij}. Hence fc this u_{ij} we cannot have $\Gamma_\omega \vdash B'$ else $\Gamma_\omega \vdash f$ by I contrary to the fact th Γ_ω is consistent (property i). That is, by induction hypothesis, B has th value F for at least the one individual u_{ij} of I and so $(x)B$ has the value as asserted by the lemma for this case.

This concludes the inductive proof of the lemma. In particular th formulas of Γ_ω all have the value T for our assignment and so are simu taneously satisfiable in the denumerable domain I. Since the formulas Λ are included among those of Γ_ω our theorem is proved for the case of system S_0 whose primitive symbols are denumerable.

To modify the proof in the case of an arbitrary system S_0 it is on necessary to replace the set of symbols u_{ij} by symbols $u_{i\alpha}$, where i rang over the positive integers as before but α ranges over a set with the san cardinal number as the set of primitive symbols of S_0; and to fix on son particular well-ordering of the formulas of the new S_ω in place of th standard enumeration employed above. (Of course the axiom of choi must be used in this connection.)

The completeness of the system S_0 is an immediate consequence of o theorem.

COROLLARY 1. *If A is a valid* wff *of S_0 then* $\vdash A$.

First consider the case where A is a cwff. Since A is valid $A \supset f$ has t value F for any assignment with respect to any domain; i.e., $A \supset f$ is n satisfiable. By our theorem it is therefore inconsistent: $A \supset f \vdash f$. Hen $\vdash A \supset f \supset f$ by III and $\vdash A$ by 3 and I.

The case of wff A' which contains some free occurrence of an individu variable may be reduced to the case of the cwff A (the *closure* of A') obtain by prefixing to A' universal quantifiers with respect to each individu variable with free occurrences in A' (in the order in which they appear). F it is clear from the definition of validity that if A' is valid so is A. But th $\vdash A$. From which we may infer $\vdash A'$ by successive applications of 5 and I

COROLLARY 2. *Let S_0 be a functional calculus of first order and* **m** *cardinal number of the set of its primitive symbols. If Λ is a set of* cw *vhich is simultaneously satisfiable then in particular Λ is simultaneously atisfiable in some domain of cardinal* **m**.

This is an immediate consequence of our theorem and the fact that if Λ is imultaneously satisfiable it must also be consistent (since rules of inference reserve the property of having the value T for any particular assignment in ny domain, and so could not lead to the formula f). For the special case vhere **m** is $\aleph_0$ this corollary is the well-known Skolem–Löwenheim result.[6] t should be noticed, for this case, that the assertion of a set of cwffs Λ can o more compel a domain to be finite than non-denumerably infinite: there s always a denumerably infinite domain available. There are also always omains of any cardinality greater than $\aleph_0$ in which a consistent set Λ is imultaneously satisfiable, and sometimes finite domains. However, for ertain Λ no finite domain will do.

Along with the truth functions of propositional calculus and quantifiation with respect to individual variables the first-order functional calculus s sometimes formulated so as to include the notion of equality as between ndividuals. Formally this may be accomplished by singling out some funconal constant of degree 2, say Q, abbreviating $Q(x,y)$ as $x=y$ (for indiidual symbols x, y), and adding the axiom schemata

E1. $x=x$.

E2. $x=y \supset . A \supset B$, where B is obtained from A by replacing some free ccurrence of x by a free occurrence of y.

For a system S_0' of this kind our theorem holds if we replace 'the same ardinal number as' by 'a cardinal number not greater than', where the efinition of 'simultaneously satisfiable' must be supplemented by the proision that the symbol '=' shall denote the relation of equality between ndividuals. To prove this we notice that a set of cwffs Λ in the system S_0' ay be regarded as a set of cwffs (Λ, E_1', E_2') in the system S_0, where E_i' is e set of closures of axioms Ei ($i = 1, 2$). Since E_1', $E_2' \vdash x=y \supset y=x$ and $_1'$, $E_2' \vdash x=y \supset . y=z \supset x=z$ we see that the assignment which gives a alue T to each formula of Λ, E_1', E_2' must assign some equivalence relation the functional symbol Q. If we take the domain I' of equivalence classes etermined by this relation over the original domain I of constants, and ssign to each individual constant (as denotation) the class determined by self, we are led to a new assignment which is easily seen to satisfy Λ imultaneously) in S_0'.

A set of wffs may be thought of as a set of axioms determining certain omains as models; namely, domains in which the wffs are simultaneously tisfiable. For a first-order calculus containing the notion of equality we

6 Th. Skolem, 'Über einige Grundlagenfragen der Mathematik', *Skrifter utgitt av et Norske Videnskaps-Akademi i Oslo*, I, 1929, no. 4.

can find axiom sets which restrict models to be finite, unlike the situation for calculi without equality. More specifically, given any finite set of finite numbers there exist axiom sets whose models are precisely those domains in which the number of individuals is one of the elements of the given set. (For example, if the set of numbers is the pair (1,3) the single axiom

$$(x)(y)(x=y) \vee \mathbin{\blacksquare} (\exists x)(\exists y)(\exists z) \mathbin{\blacksquare} \sim (x=y) \wedge \sim (x=z) \wedge \sim (y=z) \wedge (t) \mathbin{\blacksquare} t=x \vee t=y \vee t=z$$

will suffice, where $A \wedge B$, $A \vee B$ abbreviate $\sim(A \supset \sim B)$, $A \supset B \supset B$ respectively.) However, an axiom set which has models of arbitrarily large finite cardinality must also possess an infinite model as one sees by considering the formulas

$$C_i\colon \quad (\exists x_1)(\exists x_2) \ldots (\exists x_i) \mathbin{\blacksquare} \sim (x_1 = x_2) \wedge \sim (x_1 = x_3) \ldots \wedge \sim (x_{i-1} = x_i).$$

Since by hypothesis any finite number of the C_i are simultaneously satisfiable they are consistent. Hence all the C_i are consistent and so simultaneously satisfiable—which can happen only in an infinite domain of individuals.

There are axiom sets with no finite models—namely, the set of all formulas C_i defined above. Every axiom set with an infinite model has models with arbitrary infinite cardinality. For if α, β range over any set whatever the set of all formulas $\sim(x_\alpha = x_\beta)$ for distinct α, β will be consistent (since the assumption of an infinite model guarantees consistency for any finite set of these formulas) and so can be simultaneously satisfied.

In simplified form the proof of our theorem and corollary 1 may be carried out for the propositional calculus. For this system the symbols u_i and the construction of $\boldsymbol{S}_\omega$ may be omitted, an assignment of values being made directly from Γ_0. While such a proof of the completeness of the propositional calculus is short compared with other proofs in the literature the latter are to be preferred since they furnish a constructive method for finding a formal proof of any given tautology, rather than merely demonstrate its existence.[7]

[7] [Although it does not interfere with the substance of Henkin's article, it should be noted that the definition of a 'formal proof from assumptions Γ' (p. 44) is somewhat defective. For a discussion of the defect and of a method for correcting it, see Richard Montague and Leon Henkin, 'On the Definition of Formal Deduction', *Journal of Symbolic Logic*, Vol. 21 (1958), pp. 1–9. Ed.]

III

꞉OMPLETENESS IN THE THEORY OF TYPES[1]

LEON HENKIN[2]

HE first order functional calculus was proved complete by Gödel[3] in 1930. oughly speaking, this proof demonstrates that each formula of the calculus a formal theorem which becomes a true sentence under every one of a ꞉rtain intended class of interpretations of the formal system.

For the functional calculus of second order, in which predicate variables ıay be bound, a very different kind of result is known: no matter what ecursive) set of axioms are chosen, the system will contain a formula which valid but not a formal theorem. This follows from results of Gödel[4] ɔncerning systems containing a theory of natural numbers, because a nite categorical set of axioms for the positive integers can be formulated ithin a second order calculus to which a functional constant has been dded.

By a valid formula of the second order calculus is meant one which expresses a true proposition whenever the individual variables are interpreted s ranging over an (arbitrary) domain of elements while the functional ariables of degree n range over all sets of ordered n-tuples of individuals. nder this definition of validity, we must conclude from Gödel's results ıat the calculus is essentially incomplete.

It happens, however, that there is a wider class of models which furnish n interpretation for the symbolism of the calculus consistent with the usual

1 The material in this paper is included in 'The completeness of formal systems', a hesis presented to the faculty of Princeton University in candidacy for the degree of octor of Philosophy and accepted in October 1947. The results were announced at ıe meeting of the Association for Symbolic Logic in December 1947 (cf. *Journal of ymbolic Logic*, Vol. 13 (1948), p. 61).

2 The author wishes to thank Professor Alonzo Church for encouragement, sug꞉stion, and criticism in connection with the writing of his Thesis, and to acknowledge ıe aid of the National Research Council who supported that project with a preɔctoral fellowship.

3 Kurt Gödel, 'Die Vollständigkeit der Axiome des logischen Funktionenkalküls', *Ionatshefte für Mathematik und Physik*, Vol. 37 (1930), pp. 349–60.

4 Kurt Gödel, 'Über formal unentscheidbare Sätze der Principia Mathematica und ꞉rwandter Systeme I', *Monatshefte für Mathematik und Physik*, Vol. 38 (1931), pp. ⁷3–98.

axioms and formal rules of inference. Roughly, these models consist of a[n] arbitrary domain of individuals, as before, but now an *arbitrary*[5] *class* o[f] sets of ordered n-tuples of individuals as the range for functional variable[s] of degree n. If we redefine the notion of valid formula to mean one whic[h] expresses a true proposition with respect to every one of *these* models, w[e] can then prove that the usual axiom system for the second order calculus i[s] complete: a formula is valid if and only if it is a formal theorem.[6]

A similar result holds for the calculi of higher order. In this paper, we wi[ll] give the details for a system of order ω embodying a simple theory of (finite[)] types. We shall employ the rather elegant formulation of Church,[7] th[e] details of which are summarized below:

Type symbols (to be used as subscripts):

1. o and ι are type symbols.
2. If α, β are type symbols so is $(\alpha\beta)$.

Primitive symbols (where α may be any type symbol):

Variables: $f_\alpha, g_\alpha, x_\alpha, y_\alpha, z_\alpha, f'_\alpha, g'_\alpha, \ldots$
Constants: $N_{(oo)}, A_{((oo)o)}, \Pi_{(o(o\alpha))}, \iota_{(\alpha(o\alpha))}$
Improper: λ, (,)

Well-formed formulas (wffs) and their *type*:

1. A variable or constant alone is a wff and has the type of its subscript.
2. If $A_{\alpha\beta}$ and B_β are wffs of type $(\alpha\beta)$ and β respectively, then $(A_{\alpha\beta}B_\beta$[)] is a wff of type α.
3. If A_α is a wff of type α and α_β a variable of type β then $(\lambda a_\beta A_\alpha)$ is [a] wff of type $(\alpha\beta)$.

An occurrence of a variable a_β is *bound* if it is in a wff of the form $(\lambda a_\beta A_\alpha)$[;] otherwise the occurrence is *free*.

[5] These classes cannot really be taken in an altogether arbitrary manner if ever[y] formula is to have an interpretation. For example, if the formula $F(x)$ is interprete[d] as meaning that x is in the class F, then $\sim F(x)$ means that x is in the complement o[f] F; hence the range for functional variables such as F should be closed under com[-]plementation. Similarly, if G refers to a set of ordered pairs in some model, then th[e] set of individuals x satisfying the formula $(\exists y)G(x,y)$ is a projection of the set G[;] hence, we require that the various domains be closed under projection. In short, eac[h] method of compounding formulas of the calculus has associated with it some opera[-]tion on the domains of a model, with respect to which the domains must be close[d.] The statement of completeness can be given precisely and proved for models meetin[g] these closure conditions.

[6] A demonstration of this type of completeness can be carried out along the lines o[f] the author's recent paper, 'The completeness of the first order functional calculus'[,] *Journal of Symbolic Logic*, Vol. 14 (1949), pp. 159–66. [Present volume, pp. 42–50. Ed.[]]

[7] Alonzo Church, 'A formulation of the simple theory of types', *Journal of Symboli[c]* *Logic*, Vol. 5 (1940), pp. 56–68.

Letters A_α, B_α, C_α, will be used as syntactical variables for wffs of type α.

Abbreviations:

$(\sim A_o)$	for $(N_{(oo)} A_o)$
$(A_o \vee B_o)$	for $((A_{((oo)o)} A_o) B_o)$
$(A_o B_o)$	for $(\sim((\sim A_o) \vee (\sim B_o)))$
$(A_o \supset B_o)$	for $((\sim A_o) \vee B_o)$
$(a_\alpha) B_o$	for $(\Pi_{(o(o\alpha))}(\lambda a_\alpha B_o))$
$(\exists a_\alpha) B_o$	for $(\sim((a_\alpha)(\sim A_o)))$
$(\iota a_\alpha B_o)$	for $(\iota_{(\alpha(o\alpha))}(\lambda a_\alpha B_o))$
$Q_{((o\alpha)\alpha)}$	for $(\lambda x_\alpha(\lambda y_\alpha(f_{o\alpha})((f_{o\alpha} x_\alpha) \supset (f_{o\alpha} y_\alpha))))$
$(A_\alpha = B_\alpha)$	for $((Q_{((o\alpha)\alpha)} A_\alpha)(B_\alpha)$

In writing wffs and subscripts, we shall practise the omission of parentheses and their supplantation by dots on occasion, the principal rules of restoration being first that the formula shall be well-formed; secondly, that association is to the left; and thirdly, that a dot is to be replaced by a left parenthesis having its mate as far to the right as possible. (For a detailed statement of usage, refer to Church.[7])

Axioms and Axiom Schemata:

1. $(x_o \vee x_o) \supset x_o$
2. $x_o \supset (x_o \vee y_o)$
3. $(x_o \vee y_o) \supset (y_o \vee x_o)$
4. $(x_o \supset y_o) \supset . (z_o \vee x_o) \supset (z_o \vee y_o)$

5^α. $\Pi_{o(o\alpha)} f_{o\alpha} \supset f_{o\alpha} x_\alpha$

6^α. $(x_\alpha)(y_o \vee f_{o\alpha} x_\alpha) \supset . y_o \vee \Pi_{o(o\alpha)} f_{o\alpha}$

10. $x_o \equiv y_o \supset x_o = y_o \quad (x_\beta)(f_{\alpha\beta} x_\beta = g_{\alpha\beta} x_\beta) \supset f_{\alpha\beta} = g_{\alpha\beta}$

11^α. $f_{o\alpha} x_\alpha \supset f_{o\alpha}(\iota_{\alpha(o\alpha)} f_{o\alpha})$

Rules of Inference:

I. To replace any part A_α of a formula by the result of substituting a_β for b_β throughout A_α, provided that b_β is not a free variable of A_α and a_β does not occur in A_α.

II. To replace any part $(\lambda a_\gamma A_\beta) B_\gamma$ of a wff by the result of substituting B_γ for a_γ throughout A_β, provided that the bound variables of A_β are distinct both from a_γ and the free variables of B_γ.

III. To infer A_o from B_o if B_o may be inferred from A_o by a single application of Rule II.

IV. From $A_{o\alpha} a_\alpha$ to infer $A_{o\alpha} B_\alpha$ if the variable a_α is not free in $A_{o\alpha}$.

V. From $A_o \supset B_o$ and A_o to infer B_o.

VI. From $A_{o\alpha} a_\alpha$ to infer $\Pi_{o(o\alpha)} A_{o\alpha}$ provided that the variable a_α is not free in $A_{o\alpha}$.

A finite sequence of wffs each of which is an axiom or obtained from preceding elements of the sequence by a single application of one of the rules I–VI is called a *formal proof*. If A is an element of some formal proof, we write $\vdash A$ and say that A is a *formal theorem*.

This completes our description of the formal system. In order to discuss the question of its completeness, we must now give a precise account of the manner in which this formalism is to be *interpreted*.

By a *standard model*, we mean a family of domains, one for each type-symbol, as follows: D_ι is an arbitrary set of elements called *individuals*, D_o is the set consisting of two truth values, T and F, and $D_{\alpha\beta}$ is the set of all functions defined over D_β with values in D_α.

By an *assignment* with respect to a standard model $\{D_\alpha\}$, we mean a mapping ϕ of the variables of the formal system into the domains of the model such that for a variable a_α of type α as argument, the value $\phi(a_\alpha)$ of ϕ is an element of D_α.

We shall associate with each assignment ϕ a mapping V_ϕ of all the formulas of the formal system such that $V_\phi(A_\alpha)$ is an element of D_α for each wff A_α of type α. We shall define the values $V_\phi(A_\alpha)$ simultaneously for all ϕ by induction on the length of the wff A_α:

(i) If A_α is a variable, set $V_\phi(A_\alpha) = \phi(A_\alpha)$. Let $V_\phi(N_{oo})$ be the function whose values are given by the table

x	$V_\phi(N_{oo})(x)$
T	F
F	T

Let $V_\phi(A_{ooo})$ be the function whose value for arguments T, F are the functions given by the tables 1, 2 respectively.

1.

x	$V_\phi(A_{ooo})(\mathrm{T})(x)$
T	T
F	T

2.

x	$V_\phi(A_{ooo})(\mathrm{F})(x)$
T	T
F	F

Let $V_\phi(\Pi_{o(o\alpha)})$ be the function which has the value T just for the single argument which is the function mapping D_α into the constant value T. Let $V_\phi(\iota_{\alpha(o\alpha)})$ be some fixed function whose value for any argument f of $D_{o\alpha}$ is one of the elements of D_α mapped into T by f (if there is such an element).

(ii) If A_α has the form $B_{\alpha\beta} C_\beta$ define $V_\phi(B_{\alpha\beta} C_\beta)$ to be the value of the function $V_\phi(B_{\alpha\beta})$ for the argument $V_\phi(C_\beta)$.

(iii) Suppose A_α has the form $(\lambda a_\beta B_\gamma)$. We define $V_\phi(\lambda a_\beta B_\gamma)$ to be the ınction whose value for the argument x of D_β is $V_\psi(B_\gamma)$, where ψ is the ssignment which has the same values as ϕ for all variables except a_β, while (a_β) is x.

We can now define a wff A_o to be *valid in the standard sense* if $V_\phi(A_o)$ is for every assignment ϕ with respect to every standard model $\{D_\alpha\}$.[8] ecause the theory of recursive arithmetic can be developed within our rmal system as shown by Church,[7] it follows by Gödel's methods[4] that we n construct a particular wff A_o which is valid in the standard sense, but ot a formal theorem.

We can, however, interpret our formalism with respect to other than the andard models. By a *frame*, we mean a family of domains, one for each pe symbol, as follows: D_ι is an arbitrary set of individuals, D_o is the set of o truth values, T and F, and $D_{\alpha\beta}$ is some class of functions defined over $_\beta$ with values in D_α.

Given such a frame, we may consider assignments ϕ mapping variables the formal system into its domains, and attempt to define the functions $_\phi$ exactly as for standard models. For an arbitrary frame, however, it may ell happen that one of the functions described in items (i), (ii), or (iii) as the lue of some $V_\phi(A_\alpha)$ is not an element of any of the domains.

A frame such that for every assignment ϕ and wff A_α of type α, the value $_\phi(A_\alpha)$ given by rules (i), (ii), and (iii) is an element of D_α, is called a *general odel*. Since this definition is impredicative, it is not immediately clear that y non-standard models exist. However, they do exist (indeed, there are neral models for which every domain D_α is denumerable), and we shall ve a method of constructing every general model without resorting to predicative processes.

Now we define a *valid formula in the general sense* as a formula A_o such at $V_\phi(A_o)$ is T for every assignment ϕ with respect to any general model. e shall prove a completeness theorem for the formal system by showing at A_o is valid in the general sense if and only if $\vdash A$.

By a *closed* well-formed formula (cwff), we mean one in which on occur- ice of any variable is free. If Λ is a set of cwffs such that, when added to

In addition to the notion of validity, the mappings V_ϕ may be used to define the ncept of the *denotation* of a wff A_α containing no free occurrence of any variable. first show (by induction) that if ϕ and ψ are two assignments which have the same ue for every variable with a free occurrence in the wff B_α, then $V_\phi(B_\alpha) = V_\psi(B_\alpha)$. en the denotation of A_α is simply $V_\phi(A_\alpha)$ for any ϕ. We also define the notion of isfiability. If Γ is a set of wffs and ϕ an assignment with respect to some model $\{D_\alpha\}$ h that $V_\phi(A_o)$ is T for every A_o in Γ, then we say that Γ is *satisfiable with respect the model* $\{D_\alpha\}$. If Γ is satisfiable with respect to some model, we say simply that s *satisfiable*.

the axioms 1–6^{α}, 10^{α}, 11^{α}, a formal proof can be obtained for some wff A_o, we write $\Lambda \vdash A_o$. If $\Lambda \vdash A_o$ for every wff A_o, we say that Λ is *inconsistent*, otherwise *consistent*.

THEOREM 1. *If Λ is any consistent set of* cwffs, *there is a general model (in which each domain D_{α} is denumerable) with respect to which Λ is satisfiable.*

We shall make use of the following derived results about the forma calculus which we quote without proof:

VII. The deduction theorem holds: If Λ, $A_o \vdash B_o$, then $\Lambda \vdash A_o \supset B_o$ where Λ is any set of cwffs, A_o is any cwff, and B_o is any wff. (A proof is given in Church.[7])

12. $\vdash A_o \supset\centerdot \sim A_o \supset B_o$
13. $\vdash A_o \supset B_o \supset\centerdot \sim A_o \supset B_o \supset\centerdot B_o$
14. $\vdash A_{\alpha} = A_{\alpha}$
15. $\vdash A_{\alpha} = B_{\alpha} \supset B_{\alpha} = A_{\alpha}$
16. $\vdash A_{\alpha} = B_{\alpha} \supset\centerdot B_{\alpha} = C_{\alpha} \supset\centerdot A_{\alpha} = C_{\alpha}$
17. $\vdash A_o \supset\centerdot (A_o = B_o) \supset B_o$
18. $\vdash \sim A_o \supset\centerdot (A_o = B_o) \supset \sim B_o$
19. $\vdash A_o \supset\centerdot B_o \supset\centerdot A_o = B_o$
20. $\vdash \sim A_o \supset\centerdot \sim B_o \supset\centerdot A_o = B_o$
21. $\vdash A_{\alpha\beta} = A'_{\alpha\beta} \supset\centerdot B_{\beta} = B'_{\beta} \supset\centerdot A_{\alpha\beta} B_{\beta} = A'_{\alpha\beta} B'_{\beta}$
22. $\vdash A_{\alpha\beta}((\text{℩} x_{\beta}) \sim (A_{\alpha\beta} x_{\beta} = A'_{\alpha\beta} x_{\beta})) = A'_{\alpha\beta}((\text{℩} x_{\beta}) \sim (A_{\alpha\beta} x_{\beta} = A'_{\alpha\beta} x_{\beta}))$
    $\supset\centerdot A_{\alpha\beta} = A'_{\alpha\beta}$
23. $\vdash A_o \supset \sim\sim A_o$
24. $\vdash C_o \supset\centerdot C_o \vee A_o$
25. $\vdash \sim C_o \supset\centerdot A_o \supset\centerdot C_o \vee A_o$
26. $\vdash \sim C_o \supset\centerdot \sim A_o \supset\centerdot \sim (C_o \vee A_o)$
27. $\vdash \Pi_{o(o\alpha)} A_{o\alpha} \supset A_{o\alpha} C_{\alpha}$
28. $\vdash A_{o\alpha}((\text{℩} x_{\alpha}) \sim (A_o x_{\alpha})) \supset \Pi_{o(o\alpha)} A_{o\alpha}$
29. $\vdash A_{o\alpha} C_{\alpha} \supset A_{o\alpha}(\iota_{\alpha(o\alpha)} C_{\alpha})$
30. $\vdash (\sim B_o \supset B_o) \supset B_o$
31. $\vdash (x_{\alpha}) A_o \supset A_o$

The first step in our proof of Theorem 1 is to construct a maxim consistent set Γ of cwffs such that Γ contains Λ, where by maximal is mea that if A_o is any cwff not in Γ then the enlarged set $\{\Gamma, A_o\}$ is inconsister Such a set Γ may be obtained in many ways. If we enumerate all of the cw in some standard order, we may test them one at a time, adding them to and previously added formulas whenever this does not result in an inco sistent set. The union of this increasing sequence of sets is then easily se to be maximal consistent.

Γ has certain simple properties which we shall use. If A_o is any cwff, it is ear that we cannot have both $\Gamma \vdash A_o$ and $\Gamma \vdash \sim A_o$ for then by 12 and V, e would obtain $\Gamma \vdash B_o$ for any B_o, contrary to the consistency of Γ. On the her hand, at least one of the cwffs A_o, $\sim A_o$ must be in Γ. For otherwise, ing the maximal property of Γ we would have $\Gamma, A_o \vdash B_o$ and $\Gamma, \sim A_o \vdash B_o$ r any B_o. By VII, it then follows that $\Gamma \vdash A_o \supset B_o$ and $\Gamma \vdash \sim A_o \supset B_o$, ience by 13 and V $\Gamma \vdash B_o$ contrary to the consistency of Γ.

Two cwffs A_α, B_α of type α will be called *equivalent* if $\Gamma \vdash A_\alpha = B_\alpha$. Using , 16, and V, we easily see that this is a genuine congruence relation so that e set of all cwffs of type α is partitioned into disjoint equivalent classes $_\alpha$], $[B_\alpha]$, ... such that $[A_\alpha]$ and $[B_\alpha]$ are equal if and only if A_α is equivalent B_α.

We now define by induction on α a frame of domains $\{D_\alpha\}$, and simul- ieously a one–one mapping Φ of equivalence classes onto the domains $_\alpha$ such that $\Phi([A_\alpha])$ is in D_α.

D_o is the set of two truth values, T and F, and for any cwff A_o of type o $[A_o])$ is T or F according as A_o or $\sim A_o$ is in Γ. We must show that Φ is 'unction of equivalence classes and does not really depend on the par- ular representative A_o chosen. But by 17 and V, we see that if $\Gamma \vdash A_o$ and is equivalent to A_o (i.e., $\Gamma \vdash A_o = B_o$), then $\Gamma \vdash B_o$; and similarly if $\vdash A_o$ and B_o is equivalent to A_o, then $\Gamma \vdash \sim B_o$ by 18. To see that Φ is e–one, we use 19 to show that if $\Phi([A_o)]$ and $\Phi([B_o])$ are both T (i.e., $\vdash A_o$ and $\Gamma \vdash B_o$), then $\Gamma \vdash A_o = B_o$ so that $[A_o]$ is $[B_o]$. Similarly 20 ›ws that $[A_o]$ is $[B_o]$ in case $\Phi([A_o])$ and $\Phi([B_o])$ are both F.

D_ι is simply the set of equivalence classes $[A_\iota]$ of all cwffs of type ι. And $[A_\iota])$ is $[A_\iota]$ so that Φ is certainly one–one.

Now suppose that D_α and D_β have been defined, as well as the value of for all equivalence classes of formulas of type α and of type β, and that ry element of D_α, or D_β, is the value of Φ for some $[A_\alpha]$, or $[B_\beta]$ respec- ely. Define $\Phi([A_{\alpha\beta}])$ to be the function whose value, for the element $[B_\beta])$ of D_β, is $\Phi([A_{\alpha\beta}B_\beta])$. This definition is justified by the fact that if $_\beta$ and B'_β are equivalent to $A_{\alpha\beta}$ and B_β respectively, then $A'_{\alpha\beta}B'_\beta$ is iivalent to $A_{\alpha\beta}B_\beta$, as one sees by 21. To see that Φ is one–one, suppose .t $\Phi([A_{\alpha\beta}])$ and $\Phi([A'_{\alpha\beta}])$ have the same value for every $\Phi([B_\beta])$ of D_β. nce $\Phi([A_{\alpha\beta}B_\beta]) = \Phi([A'_{\alpha\beta}B_\beta])$ and so, by the induction hypothesis .t Φ is one–one for equivalence classes of formulas of type α, $A_{\alpha\beta}B_\beta$ is iivalent to $A'_{\alpha\beta}B_\beta$ for each cwff B_β. In particular, if we take B_β to be $_\beta) \sim (A_{\alpha\beta}x_\beta = A'_{\alpha\beta}x_\beta)$, we see by 22 that $A_{\alpha\beta}$ and $A'_{\alpha\beta}$ are equivalent that $[A_{\alpha\beta}] = [A'_{\alpha\beta}]$. The one–one function Φ having been thus completely ined, we define $D_{\alpha\beta}$ to be the set of values $\Phi([A_{\alpha\beta}])$ for all cwffs $A_{\alpha\beta}$.

Now let ϕ be any assignment mapping each variable x_α into some element

$\Phi([A_\alpha])$ of D_α, where A_α is a cwff. Given any wff B_β, let B^ϕ_β be a cwf obtained from B_β by replacing all free occurrences in B_β of any variable x_α by some cwff A_α such that $\phi(x_\alpha) = \Phi([A_\alpha])$.

LEMMA. *For every ϕ and B_β we have $V_\phi(B_\beta) = \Phi([B^\phi_\beta])$.*

The proof is by induction on the length of B_β.

(i) If B_β is a variable and $\phi(B_\beta)$ is the element $\Phi([A_\beta])$ of D_β, then by definition B^ϕ_β is some cwff A'_β equivalent to A_β and $V_\phi(B_\beta) = \phi(B_\beta) = \Phi([A_\beta]) = \Phi([A'_\beta]) = \Phi([B^\phi_\beta])$.

Suppose B_β is N_{oo}, whence B^ϕ_β is N_{oo}. If $\Phi([A_o])$ is T, then by definition $\Gamma \vdash A_o$ whence by 23 $\Gamma \vdash \sim N_{oo}A_o$ so that $\Phi([N_{oo}A_o])$ is F. That is $\Phi([N_{oo}])$ maps T into F. Conversely, if $\Phi([A_o])$ is F, then by definition $\Gamma \vdash N_{oo}A_o$ so that $\Phi([N_{oo}A_o])$ is T; i.e., $\Phi([N_{oo}])$ maps F into T. Hence $V_\phi(B_\beta) = \Phi([B^\phi_\beta])$ in this case.

Suppose B_β is A_{ooo}, whence B^ϕ_β is A_{ooo}. If $\Phi([C_o])$ is T, then by definition $\Gamma \vdash C_o$ whence by 24 $\Gamma \vdash A_{ooo}C_oA_o$ for any A_o so that $\Phi([A_{ooo}C_oA_o])$ T no matter whether $\Phi([A_o])$ is T or F. Similarly, using 25 and 26, we se that $\Phi([A_{ooo}C_oA_o])$ is T, or F, if $\Phi([C_o])$ is F, and $\Phi([A_o])$ is T, or respectively. Comparing this with the definition of $V_\phi(A_{ooo})$, we see th the lemma holds in this case also.

Suppose B_β is $\Pi_{o(o\alpha)}$, whence B^β_ϕ is $\Pi_{o(o\alpha)}$. If the value of $\Phi([\Pi_{o(o\alpha)}$ fo the argument $\Phi([A_{o\alpha}])$ is T, then $\Gamma \vdash \Pi_{o(o\alpha)}A_{o\alpha}$ whence by 27 $\Gamma \vdash A_{o\alpha}C$ for every cwff C_α so that $\Phi([A_{o\alpha}])$ maps every element of D_α into T. On th other hand, if $\Phi([A_{o\alpha}])$ maps every $\Phi([C_\alpha])$ into T, then we have, taking th particular case where C_α is $(\imath x_\alpha) \sim (A_{o\alpha}x_\alpha)$, $\Gamma \vdash A_{o\alpha}((\imath x_\alpha) \sim (A_{o\alpha}x_\alpha$ whence by 28 $\Gamma \vdash \Pi_{o(o\alpha)}A_{o\alpha}$. That is $\Phi([\Pi_{o(o\alpha)}])$ maps $\Phi([A_{o\alpha}])$ into The lemma holds in this case.

Suppose B_β is $\iota_{\alpha(o\alpha)}$, whence B^ϕ_β is $\iota_{\alpha(o\alpha)}$. Let $A_{o\alpha}$ be a cwff such th $\Phi([A_{o\alpha}])$ maps some $\Phi([C_\alpha])$ into T so that $\Gamma \vdash A_{o\alpha}C_\alpha$. Then by $\Gamma \vdash A_{o\alpha}(\iota_{\alpha(o\alpha)}A_{o\alpha})$ so that the value of $\Phi([\iota_{\alpha(o\alpha)}])$ for the argume $\Phi([A_{o\alpha}])$ is mapped into T by the latter. Therefore, we may take $\Phi([\iota_{\alpha(o\alpha}$ to be $V_\phi(\iota_{\alpha(o\alpha)})$.

(ii) Suppose that B_β has the form $B_{\beta\gamma}C_\gamma$. We assume (induction hyp thesis) that we have already shown $\Phi([B^\phi_{\beta\gamma}]) = V_\phi(B_{\beta\gamma})$ and $\Phi([C^\phi_\gamma])$ $V_\phi(C_\gamma)$.

Now $V_\phi(B_{\beta\gamma}C_\gamma)$ is the value of $V_\phi(B_{\beta\gamma})$ for the argument $V_\phi(C_\gamma)$, or t value of $\Phi([B^\phi_{\beta\gamma}])$ for the argument $\Phi([C^\phi_\gamma])$, which is $\Phi([B^\phi_{\beta\gamma}C^\phi_\gamma])$. B $(B_{\beta\gamma}C_\gamma)^\phi$ is simply $B^\phi_{\beta\gamma}C^\phi_\gamma$. Hence $V_\phi(B_{\beta\gamma}C_\gamma) = \Phi([B_{\beta\gamma}C_\gamma)^\phi])$.

(iii) Suppose that B_β has the form $\lambda a_\gamma C_\alpha$ and our induction hypothe is that $\Phi([C^\phi_\alpha]) = V_\phi(C_\alpha)$ for every assignment ϕ. Let $\Phi([A_\gamma])$ be any e ent of D_γ. Then the value of $\Phi([(\lambda a_\gamma C_\alpha)^\phi])$ for the argument $\Phi([A_\gamma])$ is by efinition $\Phi([(\lambda a_\gamma C_\alpha)^\phi A_\gamma])$.

But by applying II to the right member of the instance $\vdash (\lambda a_\gamma C_\alpha)^\phi A_\gamma =$ $a_\gamma C_\alpha)^\phi A_\gamma$ of 14, we find $\vdash (\lambda a_\gamma C_\alpha)^\phi A_\gamma = C_\alpha^\psi$, where ψ is the assignment hich has the same value as ϕ for every argument except the variable a_γ and (a_γ) is $\Phi([A_\gamma])$. That is, $[(\lambda a_\gamma C_\alpha)^\phi A_\gamma] = [C_\alpha^\psi]$ so that the value of $([(\lambda a_\gamma C_\alpha)^\phi])$ for the argument $\Phi([A_\gamma])$ is $\Phi([C_\alpha^\psi])$—or $V_\psi(C_\alpha)$ by induc-on hypothesis. Since for every argument $\Phi([(\lambda a_\gamma C_\alpha)^\phi])$ and $V_\phi(\lambda a_\gamma C_\alpha)$ ave the same value, they must be equal.

This concludes the proof of our lemma.

Theorem 1 now follows directly from the lemma. In the first place, the ame of domains $\{D_\alpha\}$ is a general model since $V_\phi(B_\beta)$ is an element of D_β r every wff B_β and assignment ϕ. Because the elements of any D_α are in ne–one correspondence with equivalence classes of wffs each domain is enumerable. Since for every cwff $A_o^\phi = A_o$, ϕ being an arbitrary assign-ent, since therefore for every cwff A_o of Γ we have $\Phi([A_o]) = \mathrm{T}$, and since is a subset of Γ, it follows that $V_\phi(A_o)$ is T for any element A_o of Λ; i.e., is satisfiable with respect to the model $\{D_\alpha\}$.

THEOREM 2. *For any* wff A_o, *we have* $\vdash A_o$ *if and only if* A_o *is valid in the eneral sense.*

From the definition of validity, we easily see that A_o is valid if and only the cwff $(x_{\alpha 1}) \ldots (x_{\alpha n}) A_o$ is valid, where $x_{\alpha 1}, \ldots, x_{\alpha n}$ are the variables with ee occurrences in A_o; and hence A_o is valid if and only if $V_\phi(\sim(x_{\alpha 1}) \ldots$ $_{\alpha n}) A_o)$ is F for every assignment ϕ with respect to every general model $_\alpha\}$. By Theorem 1, this condition implies that the set Δ whose only ele-ent is the cwff $\sim(x_{\alpha 1}) \ldots (x_{\alpha n}) A_o$ is inconsistent and hence, in particular, $(x_{\alpha 1}) \ldots (x_{\alpha n}) A_o \vdash (x_{\alpha 1}) \ldots (x_{\alpha n}) A_o$. Now applying VII, 30, and 31 everal times), we see that if A_o is valid, then $\vdash A_o$. The converse can be rified directly by checking the validity of the axioms and noticing that the les of inference operating on valid formulas lead only to valid formulas.

THEOREM 3. *A set* Γ *of* cwffs *is satisfiable with respect to some model of numerable domains* D_α *if and only if every finite subset* Λ *of* Γ *is satisfiable.*

By Theorem 1, if Γ is not satisfiable with respect to some model of enumerable domains, then Γ is inconsistent so that, in particular, $\vdash (x_o) x_o$. Since the formal proof of $(x_o) x_o$ contains only a finite number formulas, there must be some finite subset $\Lambda = \{A_1, \ldots, A_n\}$ of Γ such at $A_1, \ldots, A_n \vdash (x_o) x_o$, whence by repeated applications of VII, $\vdash A_1 \supset$. $\supset$. $A_n \supset (x_o) x_o$. But then by Theorem 2, the cwff $A_1 \supset \ldots A_n \supset (x_o) x_o$ valid so that we must have some $V_\phi(A_i) = \mathrm{F}$, $i = 1, \ldots, n$, for any ϕ with spect to any model; i.e., Λ is not satisfiable. Thus, if every finite subset Λ

of Γ is satisfiable, then Γ is satisfiable with respect to a model of denumerable domains. The converse is immediate.

If Γ is satisfiable, then so are its finite subsets, and hence Γ is satisfiable with respect to some model of denumerable domains. This may be taken as a generalization of the Skolem–Löwenheim theorem for the first order functional calculus.

Analogues of Theorems 1, 2, and 3 can be proved for various formal systems which differ in one respect or another from the system which we have here considered in detail. In the first place, we may add an arbitrary set of *constants* S_α as new primitive symbols. In case the set of constants is infinite, we must replace the condition of denumerability, in the statement of Theorems 1 and 3, by the condition that the domains of the model will have a cardinality not greater than that of the set of constants. The proofs for such systems are exactly like the ones given here.

In the second place, the symbols $\iota_{\alpha(o\alpha)}$ and the axioms of choice (11^α) may be dropped. In this case, we have to complicate the proof by first performing a construction which involves forming a sequence of formal systems built up from the given one by adjoining certain constants u_α^{ij}, $i, j = 1, 2, \ldots$, and providing suitable axioms for them. The details can be obtained by consulting the paper mentioned in footnote 6.

The axioms of extensionality (10^α) can be dropped if we are willing to admit models whose domains contain functions which are regarded as distinct even though they have the same value for every argument.

Finally, the functional abstraction of the present system may either be replaced by set-abstraction or dropped altogether. In the latter case, the constants $\Pi_{o(o\alpha)}$ must be replaced by a primitive notion of quantifiers.

Theorem 3 can be applied to throw light on formalized systems of number theory.

The concepts of elementary number theory may be introduced into the pure functional calculus of order ω by definition, a form particularly suited to the present formulation being given in Church.[7] Under this approach, the natural numbers are identified with certain functions. Alternatively we may choose to identify the natural numbers with the individuals making up the domain of type ι. In such a system, it is convenient to construct an applied calculus by introducing the constants 0_ι and $S_{\iota\iota}$ and adding the following formal equivalents of Peano's postulates:

P1. $(x_\iota) \centerdot \sim S_{\iota\iota} x_\iota = 0_\iota$

P2. $(x_\iota)(y_\iota) \centerdot S_{\iota\iota} x_\iota = S_{\iota\iota} y_\iota \supset x_\iota = y_\iota$

P3. $(f_{o\iota}) \centerdot f_{o\iota} 0_\iota \supset \centerdot (x_\iota)[f_{o\iota} x_\iota \supset f_{o\iota}(S_{\iota\iota} x_\iota)] \supset (x_\iota) f_{o\iota} x_\iota$

The Peano axioms are generally thought to characterize the number-
:quence fully in the sense that they form a categorical axiom set any two
odels for which are isomorphic. As Skolem[9] points out, however, this
ondition obtains only if 'set'—as it appears in the axiom of complete
iduction (our P3)—is interpreted with its standard meaning. Since, how-
ver, the scope ('all sets of individuals') of the quantifier $(f_{o\iota})$ may vary
om one general model to another,[10] it follows that we may expect non-
andard models for the Peano axioms.

This argument may be somewhat clearer if we consider in detail the usual
oof of the categoricity of Peano's postulates. One easily shows that any
odel for the axioms must *contain* a sequence of the order-type of the
atural numbers by considering the individuals 0_ι, $S_{\iota\iota}0_\iota$, $S_{\iota\iota}(S_{\iota\iota}0_\iota)$, ... and
sing P1 and P2 to show them distinct and without other predecessors. Then
ie proof continues as follows.

Suppose that the domain of individuals contained elements other than
iose of this sequence (which we may as well identify with the natural
umbers themselves). Then consider the class of individuals consisting of
st the natural numbers. Since it contains zero (0_ι) and is closed under the
ccessor function $(S_{\iota\iota})$, we infer from the axiom of complete induction
'3) that it contains all individuals, contrary to the hypothesis that some
dividuals were not numbers.

By examining this proof, we see that we can conclude only that if a general
odel satisfies Peano's axioms and at the same time possesses a domain of
dividuals not isomorphic to the natural numbers, then the domain $D_{o\iota}$ of
ts of individuals *cannot* contain the set consisting of just those individuals
hich are numbers.

Although Skolem indicates that the meaning of 'natural number' is rela-
ve to the variable meaning of 'set' he does not give any example of a
on-standard number system satisfying all of Peano's axioms. In two later
ipers,[11] however, he proves that it is impossible to characterize the natural
imber sequence by any denumerable system of axioms formulated within
e first order functional calculus (to which may be added any set of funconal constants denoting numerical functions and relations), the individual

9 Thoralf Skolem, 'Über einige Grundlagenfragen der Mathematik', *Skrifter utgitt Det Norske Videnskaps-Akademi*, I, no. 4 (1929), 49 pp.

10 Here we are identifying a set X of elements of D_ι with the function (element of $_{o\iota}$) which maps every element of X into T and every other element of D_ι into F.

11 Thoralf Skolem, 'Über die Unmöglichkeit einer vollständigen Charakterisierung r Zahlenreihe mittels eines endlichen Axiomensystems', *Norsk matematisk forenings rifter*—series 2, no. 10 (1933), pp. 73–82. And 'Über die Nicht-charakterisierbarkeit r Zahlenreihe mittels endlich oder abzählbar unendlich vieler Aussagen mit aus-nliesslich Zahlenvariablen', *Fundamenta mathematicae*, Vol. 23 (1934), pp. 150–61.

variables ranging over the 'numbers' themselves. Skolem makes ingeniou use of a theorem on sequences of functions (which he had previously proved to construct, for each set of axioms for the number sequence (of the typ described above) a set of numerical functions which satisfy the axioms, bu have a different order type than the natural numbers. This result, for axion systems which do not involve class variables, cannot be regarded as being a all paradoxical since the claim had never been made that such systems wer categorical.

By appealing to Theorem 3, however, it becomes a simple matter t construct a model containing a non-standard number system which wil satisfy all of the Peano postulates as well as any preassigned set of furthe axioms (which may include constants for special functions as well as con stants and variables of higher type). We have only to adjoin a new primitiv constant u_ι and add to the given set of axioms the infinite list of formula $u_\iota \neq 0_\iota$, $u_\iota \neq S_{\iota\iota}0_\iota$, $u_\iota \neq S_{\iota\iota}(S_{\iota\iota}0_\iota)$, ... Since any finite subset of the en larged system of formulas is clearly satisfiable, it follows from theorem tha some denumerable model satisfies the full set of formulas, and such a mode has the properties sought. By adding a non-denumerable number of primi tive constants v_ι^ξ together with all formulas $v_\iota^{\xi_1} \neq v_\iota^{\xi_2}$ for $\xi_1 \neq \xi_2$, we ma even build models for which the Peano axioms are valid and which contai a number system having any given cardinal.[12]

These same remarks suffice to show more generally that no mathematic axiom system can be genuinely categorical (determine its models to withi isomorphism) unless it constrains its domain of elements to have som definite finite cardinal number—provided that the logical notions of set an function are axiomatized along with the specific mathematical notions.

The existence of non-standard models satisfying axiom-systems for num ber theory throws new light on the phenomenon of ω-inconsistency, fir investigated by Tarski and Gödel. A formal system is ω-inconsistent if fc some formula $A_{o\iota}$ the formulas $A_{o\iota}0_\iota$, $A_{o\iota}(S_{\iota\iota}0_\iota)$, $A_{o\iota}(S_{\iota\iota}(S_{\iota\iota}0_\iota))$, .. $\sim(x_\iota)A_{o\iota}x_\iota$ are all provable. Tarski, and later Gödel, showed the existenc of consistent systems which were ω-inconsistent. We can now see that suc systems can and must be interpreted as referring to a non-standard numb system whose elements include the natural numbers as a proper subset.

It is generally recognized that all theorems of number theory now in th literature can be formalized and proved within the functional calculus c

12 A similar result for formulations of arithmetic within the first order function calculus was established by A. Malcev, 'Untersuchungen aus dem Gebiete der math matischen Logik', *Recueil mathématique*, n.s. Vol. 1 (1936), pp. 323–36. Malce method of proof bears a certain resemblance to the method used above. I am indebt to Professor Church for bringing this paper to my attention. (Added February 1 1950).

rder ω with axioms P1–P3 added. (In fact, much weaker systems suffice.) n the one hand, it follows from Theorem 1 that these systems can be einterpreted as true assertions about a great variety of number-systems ther than the natural numbers. On the other hand, it follows from the sults of Gödel[4] that there are true theorems about the natural numbers hich cannot be proved by extant methods (consistency assumed).

Now Gödel's proof furnishes certain special formulas which are shown to e true but unprovable, but there is no general method indicated for estab- shing that a given theorem cannot be proved from given axioms. From heorem 1, we see that such a method is supplied by the procedure of onstructing non-standard models for number theory in which 'set' and unction' are reinterpreted. It, therefore, becomes of practical interest to umber-theorists to study the structure of such models.

A detailed investigation of these numerical structures is beyond the scope the present paper. As an example, however, we quote one simple result: very non-standard denumerable model for the Peano axioms has the order pe $\omega + (\omega^* + \omega)\eta$, where η is the type of the rationals.

IV

LANGUAGES IN WHICH SELF REFERENCE IS POSSIBLE[1]

RAYMOND M. SMULLYAN

1. *Introduction.* This paper treats of semantical systems S of sufficier strength so that for any set W definable in S (in a sense which will be mad precise), there must exist a sentence X which is true in S if and only if it an element of W.[2] We call such an X a *Tarski* sentence for W. It is th sentence which (in a purely extensional sense) says of itself that it is in W If W is the set of all expressions not provable in some syntactical system (then X is the Gödel sentence which is true (in S) if and only if it is nc provable (in C). We provide a novel method for the construction of the sentences, which yields sentences particularly simple in structure. Th method is applicable to a variety of systems, including a form of elementa arithmetic, and some systems of protosyntax self applied.[4] In application the former, we obtain an extremely simple and direct proof of a theoren which is essentially Tarski's theorem that the truth set of elementary arit metic is not arithmetically definable.

The crux of our method is in the use of a certain function, the 'norı function, which replaces the classical use of the *diagonal* function. To give

From the *Journal of Symbolic Logic*, Vol. 22 (1957), pp. 55–67. Copyright © 19 Reprinted by permission of the publishers, the American Mathematical Society, a the author.

[1] I wish to express my deepest thanks to Professor Rudolph Carnap, of the Univ sity of California at Los Angeles, and to Professor John Kemeny and Dr. Edward Cogan, of Dartmouth College, for some valuable suggestions. I also wish to tha the referee for some very helpful revisions.

[2] By a semantical system S, we mean a set E of *expressions* (strings of signs), toget with a subset S of expressions called *sentences* of S (determined by a set of rules formation), together with a subset T of S, of elements called *true* sentences of (determined by a set of 'rules of truth' for S).

[3] (At the referee's suggestion)—In an extensional system, the only way we c translate the meta-linguistic phrase 'X says that $X \in W$' is by the phrase 'X is t if and only if $X \in W$'. Thus the requirement for X to be a Tarski sentence for W, exceedingly weak; any sentence X which is either both true and in W, or false a not in W, will serve. However, this is as much as we need of a Tarski sentence (undecidability results). If we were considering an *intensional* system, then we wo define a Tarski sentence for W as a sentence X which not only is true if and onl $X \in W$, but which actually expresses the proposition that $X \in W$.

[4] The former will be carried out in this paper and the latter in a forthcoming par 'Systems of Protosyntax Self Applied'.

uristic idea of the norm function, let us define the norm of an expression (of informal English) as E followed by its own quotation. Now, given a set (of expressions), to construct a sentence X which says of itself that it is W, we do so as follows:

W contains the norm of 'W contains the norm of.' This sentence X says at the norm of the expression 'W contains the norm of' is in W. However, norm of this expression is X itself. Hence X is true if and only if $X \in W$.[5]

This construction is much like one due to Quine.[6] We carry it out for ne formalized languages. In Section 2, which is essentially expository, we nstruct a very precise, though quite trivial semantical system S_P, which tes quotation and the norm function as primitive. The study of this tem will have a good deal of heuristic value, inasmuch as S_P, despite its viality, embodies the crucial ideas behind undecidability results for deeper n-trivial systems. We then consider, in Section 3, the general use of the rm function, and we finally apply the results, in Section 4, to a system S_A, ich is a formal variant of elementary arithmetic. This variant consists of ing the lower functional calculus with class abstractors, rather than antifiers, as primitive. This alteration, though in no way affecting the ength of the system, nevertheless makes possible the particularly simple oof of Tarski's or Gödel's theorem, since the arithmetization of substiion can thereby be circumvented quite simply.

3y the norm of an expression E (of S_A) we mean E followed by its own del *numeral* (i.e., the numeral designating its Gödel number). Now, given set W of expressions whose set of Gödel numbers is arithmetically inable, we show quite easily the existence of an expression H of class traction, such that for any expression E, H followed by the Gödel *neral* of E is a true sentence if and only if the *norm* of E is in W. Then, if follow H by its own Gödel numeral h, the resulting sentence Hh (which he norm of H) is true if and only if it is in W. This is a rough sketch of procedure.

. *The Preliminary System* S_0 *and the Semantical System* S_P. In this tion, we formalize the ideas behind the preceding heuristic account of norm function. For convenience, we first construct a preliminary system

In contrast with this construction, let us define the *diagonalization* of E as the lt of substituting the quotation of E for all occurrences of the variable 'x' in E. n the following Tarski sentence for W (when formalized) is the classical construc- : W contains the diagonalization of 'W contains the diagonalization of x'. This r construction involves *substitution* (inherent in diagonalization), whereas the n function involves *concatenation* (the norm of E being E *followed by* its quotation), ch is far easier to formalize (cf. Section 5).

'Yields falsehood when appended to its quotation' yields falsehood when appended s quotation. This is Quine's version of the famous semantical paradox.

S_0, whose expressions will be built from the three signs 'ϕ', '*', and 'N'. Tl second sign will serve as our formal quotation mark, since we reser ordinary quotation marks for meta-linguistic use. The sign 'N' will l endowed with the same meaning as 'the norm of.' The sign 'ϕ' will be a undefined predicate constant. For any property (set) P of expressions of S we then define the semantical system S_P by giving a rule of truth for S_P. F any P, 'ϕ' will be interpreted in S_P as designating P.

Signs of S_0: ϕ, *, N.

Preliminary Definitions. (1) By an *expression* (of S_0) we mean any stri built from the three signs of S_0. (2) By the (formal) quotation of an e pression, we mean the expression surrounded by stars. (3) By the *norm* of a expression, we mean the expression followed by its own (formal) quotatio

Formation Rules for (Individual) Designators:

(1) The quotation of any expression is a designator.

(2) If E is a designator, so is ⌜NE⌝ (i.e., 'N' followed by E).

Alternative Definition:

(1)′ A designator is an expression which is either a quotation (of sor other expression) or a quotation preceded by one or more 'N's.

Rules of Designation in S_0:

R1. The quotation of an expression E designates E.

R2. If E_1 designates E_2, then ⌜NE_1⌝ designates the *norm* of E_2.

Definition of a Sentence of S_0:

(1) A sentence of S_0 is an expression consisting of 'ϕ' followed by designator.

The Semantical System S_P:

For any property P, we define the semantical system S_P as follows:

(1) The rules for designators, designation and sentence formation in are the same as in S_0.

(2) The rule of truth for S_P is the following:

R3. For any designator E, ⌜ϕE⌝ is true in $S_P \underset{\text{df}}{=}$ the expression designa by E (in S_P) has the property P.

THEOREM 2.1. *There exists an expression of S_0, which designates itself.*

PROOF. '*N*' designates 'N' (by Rule 1).

Hence 'N*N*' designates the norm of 'N' (by Rule 2) which is 'N*N' Thus 'N*N*' designates itself.

THEOREM 2.2. *There exists a sentence G of S_0 such that for any property P,*
; is true in $S_P \Leftrightarrow$ G has the property P.

PROOF. 'N*ϕN*' designates 'ϕN*ϕN*' (by R1 and R2).
Thus *G*, viz., 'ϕN*ϕN*' is our desired sentence.

REMARK. *G* is, of course, the formalized version of '*W* contains the norm f '*W* contains the norm of'.' 'ϕ' is but an abbreviation of '*W* contains,' and
N' abbreviates 'the norm of.'[7]

COROLLARY 2.3. *P cannot be coextensive with the set of all false (non-true)* *ntences of S_P, nor is P coextensive with the set of all expressions of S_P which* *re not true sentences of S_P.*

2.4. A STRONGER FORM OF THEOREM 2.2. By a *predicate* we mean either 'ϕ' r 'ϕ' followed by one or more 'N's'.

We say that an expression E *satisfies* a predicate H (in S_P) if H followed y the quotation ⌜*E*⌝ of E, is true in S_P. Lastly, we say that a set *W* of xpressions of S_0 is *definable* (in S_P) if there exists a predicate H which is ıtisfied by all and only those expressions which are in *W*.

It is worth noting at this point, that if E_1 designates E, then E satisfies H ' and only if ⌜HE_1⌝ is true. This follows from R3 by induction on the umber of N's occurring in H.

For any set *W*, we let $\eta(W) \underset{\text{df}}{=}$ set of all expressions whose norm is in *W*.

LEMMA 2.5. *If W is definable in S_P, then so is $\eta(W)$.*

PROOF. Let H be the predicate which defines *W* (i.e., which is satisfied by ıst those elements which are in *W*). Then H followed by 'N' will be satisfied y precisely those elements which are in $\eta(W)$. Thus $\eta(W)$ is definable (in $_P$).

We can now state the following theorem, of which Theorem 2.2 is a ɔecial case.

THEOREM 2.6. *For* any *set W definable in S_P, there is a sentence X which is* *'ue in S_P if and only if $X \in W$.*

[7] If we wished to construct a miniature system L_P which formalizes the diagonal ınction in the same way as S_P does the norm function, we take four signs, viz., 'ϕ', ', '*D*', 'x', and the rules R_1, (same as S_P), R_2: If E_1 designates E_2, then ⌜DE_1⌝ esignates the diagonalization of E_2 (i.e., the result of replacing each occurrence of ' in E_2 by the quotation of E_2), R_3: If E_1 designates E_2, then ⌜ϕE_1⌝ is a sentence ıd is true in L_P if and only if E_2 has the property *P*. Then the expression of Theorem 1, which designates itself, is 'D*Dx*', and the Tarski sentence (of Theorem 2.2) for is 'ϕD*ϕDx*'.

PROOF. Assume W is definable. Then so is $\eta(W)$ (by Lemma). Hence there exists a predicate H such that for any expression E, $\ulcorner H^* E^* \urcorner$ is true (in S_P) $\Leftrightarrow E \in \eta(W)$

$$\Leftrightarrow \ulcorner E^* E^* \urcorner \in W.$$

Taking E = H, $\ulcorner H^* H^* \urcorner$ is true $\Leftrightarrow \ulcorner H^* H^* \urcorner \in W$.
Thus X, viz., $\ulcorner H^* H^* \urcorner$, is our desired sentence.

REMARK. Theorem 2.6 says (in view of the truth functionality of the biconditional) no more nor less than this: each set definable in S_P eithe contains some truths of S_P or lacks some falsehoods.

COROLLARY 2.7. *The set of false sentences of S_P is not definable in S_P, no is the complement (relative to the set of all* expressions *of S_P) of the set of tru sentences of S_P definable in S_P.*

COROLLARY 2.8. *Suppose we extend S_P to the enlarged semantical syste S'_P by adding the new sign '~', and adding the following two rules:*

R4. If X is a sentence, so is $\ulcorner \sim X \urcorner$.
R5. $\ulcorner \sim X \urcorner$ is true in $S'_P \Leftrightarrow X$ is not true in S'_P.

Then in this system S'_P, the truth set of S'_P is not definable.

PROOF. For S'_P has the property that the complement of any set definabl in S'_P is again definable in S'_P, since if H defines W, then $\ulcorner \sim H \urcorner$ defines th complement of W. Hence the truth set if not definable, since its complemen is not definable by Corollary 2.7.

REMARK. S'_P is about as simple a system as can be constructed which ha the interesting property that the truth set of the system is not definab within the system and that, moreover, any possible extension of S'_P wi retain this feature. By an extension, we mean any system constructed fror S'_P by possibly adding additional signs, and rules, but retaining the old rule in which, however, the word 'expression' is re-interpreted to mean an e pression of the enlarged system. Likewise, if we take any extension of S_P then, although we may greatly enlarge the collection of definable sets, non of them can possibly be co-extensive with the set of false sentences of th extension.

2.8. EXTENSION OF S_P TO A SEMANTICO-SYNTACTICAL SYSTEM S^C_P. Suppos now that we select an arbitrary set of sentences of S_0 and call them *axiom* and select a set of rules for inferring sentences from other sentences (c finite sets of sentences). The axioms, together with the rules of inferenc form a so-called syntactical system, or calculus C. Let S^C_P be the ordere pair (S_P, C). Thus S_P^C is a mathematical system, or interpreted calculus. We let T be the set of true sentences of S_P (also called true sentences of S_P^C) and Th, the set of sentences provable in C (also called provable sentences, or theorems, of S_P^C). We already know that the complement $\overline{\mathrm{T}}$ of T (relative to the set of expressions) is not semantically definable in S_P^C (i.e., not definable in S_P); however $\overline{\mathrm{Th}}$ may well happen to be. If it is, however, then we have, as an immediate corollary of 2.6, the following miniature version of Gödel's theorem:

THEOREM 2.9. *If the set* $\overline{\mathrm{Th}}$ *is semantically definable in* S_P^C, *then either some sentence true in* S_P^C *cannot be proved in* S_P^C *or some false sentence can be proved.*

This situation is sometimes described by saying that S_P^C is either semantically incomplete or semantically inconsistent.

2.10. We can easily construct a system S_P^C obeying the hypothesis of theorem 2.9 as follows: Before we choose a property P, we *first* construct a completely arbitrary calculus C. Then we simply define P to be the set of all expressions not provable in C. Then 'ϕ' itself will be the predicate which semantically defines $\overline{\mathrm{Th}}$ in S_P^C, and the sentence G, viz., 'ϕN*ϕN*', of Theorem 2.2 will be our Gödel sentence for S_P^C, which is true if and only if not provable in the system. In fact, for purposes of illustration, let us consider a calculus C with only a finite number of axioms, and no rules of inference. Thus the theorems of C are the axioms of C. Now, if G was included as one of the axioms, it is automatically false (in this system), whereas if G was left out, then it is true, by very virtue of being left out. Thus, this system is, with dramatic clarity, obviously inconsistent or incomplete.

REMARK. Suppose that we take P to be the set of sentences which *are* provable in C. Then G becomes the Henkin sentence for the system S_P^C, which is true in this system, if and only if G *is* provable in S_P^C. Is G true in S_P^C? This obviously depends on C. If, for example, we take C such that its set of axioms is null, then G is certainly both false and non-provable. An example of a choice of C (other than an obvious one in which G itself is an axiom) for which G is true is the following: We take for our single axiom A_1, the expression 'ϕ*ϕN*ϕN**'. We take a single rule R: If two designators E_1 and E_2 have the same designatum in S_0, then $\ulcorner\phi E_2\urcorner$ is directly derivable from $\ulcorner\phi E_1\urcorner$. This rule is 'reasonable' in the sense that it does preserve truth in S_P.

Now, is G, viz., 'ϕN*ϕN*', true in this S_P^C or not? It is true, providing it is provable. Now since 'N*ϕN*' and '*ϕN*ϕN**' both have the same designatum 'ϕN*ϕN*', then 'ϕN*ϕN*' is immediately derivable from

'$\phi^{*}\phi N^{*}\phi N^{**}$' by R, i.e., G is immediately derivable from A_1, hence G is provable, and hence also true.

We now consider whether or not this system $\boldsymbol{S}_P^C$ is semantically consistent. We have already observed that rule R does preserve truth, so the question reduces to whether or not A_1 is true. Well, by R3 of $\boldsymbol{S}_P$, A_1 is true precisely in case '$\phi N^{*}\phi N^{*}$' has the property P, i.e., precisely in case G is provable, which it is.

3. *Semantical Systems with Predicates and Individual Constants.* In this section, we consider any semantical system $\boldsymbol{S}$, of which certain expressions called *predicates* and certain expressions called (individual) *constants* are so related that any predicate followed by any constant is a sentence of $\boldsymbol{S}$. We also wish to have something of the nature of a *Gödel correspondence g* which will assign a unique constant $g(E)$ to each expression E so that $g(E)$ will be peculiar to E—i.e., we consider a 1—1 correspondence g whose domain is the set of all expressions, and whose range is a subset (proper or otherwise) of the individual constants. We shall often write '$\mathring{E}$' for '$g(E)$'. By the *norm* of E, we mean $E\mathring{E}$ (i.e., E followed by $g(E)$).[8,9] For a set W of expressions of $\boldsymbol{S}$, by $\eta(W)$ we mean the set of all E whose norm is in W.

A predicate H is said to define the set W (relative to g, understood) if W consists of all and only those expressions E such that HE is true. The following theorem, though quite simple, is basic.

THEOREM 3.1. *For any set W of expressions of $\boldsymbol{S}$, a sufficient condition for the existence of a Tarski sentence for W is that $\eta(W)$ be definable.*

PROOF. Suppose $\eta(W)$ is definable. Then for some predicate H and for any E $H\mathring{E}$ is true $\Leftrightarrow E \in \eta(W)$

$$\Leftrightarrow E\mathring{E} \in W.$$

Hence $H\mathring{H}$ is true $\Leftrightarrow H\mathring{H} \in W$.

COROLLARY 3.2. *Letting* F *be the set of non-true sentences of $\boldsymbol{S}$, and $\bar{T}$ the set of all expressions which are not true sentences, then neither $\eta(\bar{T})$ nor $\eta(F)$ are definable in $\boldsymbol{S}$.*

[8] The word 'norm' was suggested by the following usage: In Mathematics, when we have a function $f(x, y)$ of two arguments, the entity $f(x, x)$ is sometimes referred to as the norm of x—e.g., in Algebra the norm of a vector ϵ is $f(\epsilon, \epsilon)$, where f is the function which assigns to any pair of vectors the square root of their inner product. In this paper, the crucial function f (which allows us to introduce a notion of representability) is the function which assigns, to each pair (E_1, E_2) of expressions, the expression $E_1\mathring{E}_2$. Thus for this f, $f(E, E)$ is our norm of E.

[9] Professor Quine has kindly suggested that I remark that the norm of a predicate is a sentence which says in effect that the predicate is *autological*, in the sense of Grelling (i.e., that the predicate applies to itself).

COROLLARY 3.3. *If we extend* $\boldsymbol{S}$ *to a semantico-syntactical system* $\boldsymbol{S}^C$, *and* $\eta(\overline{\mathrm{Th}})$ *are definable in* $\boldsymbol{S}$, *then* $\boldsymbol{S}^C$ *is semantically incomplete or inconsistent.*

Actually, to apply Corollary 3.3 to concrete situations, one would most kely show that $\eta(\overline{\mathrm{Th}})$ is definable by showing (1) $\overline{\mathrm{Th}}$ is definable and (2) or any set W, if W is definable, so is $\eta(W)$. Semantical systems strong nough to enjoy property (2) (which is purely a property of $\boldsymbol{S}$, rather than f C) are of particular importance. We shall henceforth refer to such systems s semantically *normal* (or, more briefly, 'normal'). Thus $\boldsymbol{S}$ is normal if, henever W is definable in $\boldsymbol{S}$, so is $\eta(W)$. Semantical normality is, of course, elative to the Gödel correspondence g.

COROLLARY 3.4. *If* $\boldsymbol{S}$ *is semantically normal, then*

(1) *There is a Tarski sentence* X *for each definable set.*

(2) F *is not definable in* $\boldsymbol{S}$, *nor is* $\bar{\mathrm{T}}$.

(3) *If non-theoremhood of* C *is definable in* $\boldsymbol{S}$, *then* $\boldsymbol{S}^C$ *is semantically complete or inconsistent.*

REMARK. The trivial systems $\boldsymbol{S}_P$ of Section 2 are semantically normal elative to the correspondence g mapping each expression onto its quotation he individual constants of $\boldsymbol{S}_P$ are, of course, the designators). In fact, emma 2.5 asserts precisely that. It is by virtue of normality that we showed e non-definability of $\bar{\mathrm{T}}$ in $\boldsymbol{S}_P$. $\boldsymbol{S}_P$ was deliberately constructed with the iew of establishing normality as simply as possible.

We now turn to a non-trivial system $\boldsymbol{S}_A$, for which we easily establish mantical normality.

4. *Systems of Arithmetic*. The first arithmetical system $\boldsymbol{S}_A$ which we conder is much like arithmetic in the first order functional calculus. We have umerals (names of numbers), numerical variables, the logical connectives ll definable from the primitive '$\downarrow$' of joint denial), identity, and the primive arithmetical operations of $\cdot$ (multiplication) and $\urcorner$ (exponentiation). Ve depart from the lower functional calculus in that, given a (well-formed) rmula F and a variable, e.g., 'x', we form the (class) abstract $\ulcorner$x(F)$\urcorner$, read he set of x's such that F.' We use abstracts to form new formulas in two ays, viz., (1) For a numeral N, $\ulcorner$x(F)N$\urcorner$ (read 'N' is a member of the set f x's such that F,' or 'the set of x's such that F contains N') and (2) x(F_1) = x(F_2)$\urcorner$ (read 'the set of x's such that F_1 is identical with the set of 's such that F_2.') By (2) we easily define universal quantification thus: ($\forall$x)(F)$\urcorner$ $\underset{\mathrm{df}}{=}$ $\ulcorner$x(F) = x(x = x)$\urcorner$.

A formal description of $\boldsymbol{S}_A$ now follows.

Signs of $\boldsymbol{S}_A$: x, ′, (,), $\cdot$, $\urcorner$, =, $\downarrow$, 1. We call these signs 'S_1', 'S_2', ..., 'S_9', spectively.

Rules of Formation, Designation and Truth:

1. A numeral (string of '1's) of length n, designates the positive integer n.
2. 'x' alone, or followed by a string of accents, is a variable.
3. Every numeral and every variable is a term.
4. If t_1 and t_2 are terms, so are $\ulcorner(t_1)\cdot(t_2)\urcorner$ and $\ulcorner(t_1)\rceil(t_2)\urcorner$. If t_1 and t_2 contain no variables and respectively designate n_1 and n_2, then the above new terms respectively designate $n_1 \times n_2$ and $n_1^{n_2}$.
5. If t_1 and t_2 are terms, then $\ulcorner t_1 = t_2\urcorner$ is a formula, called an atomic formula. All occurrences of variables are free. If no variables are present, then $\ulcorner t_1 = t_2\urcorner$ is a sentence, and is a true sentence if and only if t_1 and t_2 designate the same number n.
6. If F is a formula, α a variable, then $\ulcorner\alpha(F)\urcorner$ is a (class) abstract. No occurrence of α in $\ulcorner\alpha(F)\urcorner$ is free. If β is a variable distinct from α, then the free occurrences of β in $\ulcorner\alpha(F)\urcorner$ are those in F. If F contains no free variable other than α, then $\ulcorner\alpha(F)\urcorner$ is called a *predicate*, and the *abstraction* of F.
7. If H_1 and H_2 are abstracts, then $\ulcorner H_1 = H_2\urcorner$ is a formula. The free occurrences of any variable α in this formula are those of H_1 and those of H_2.
8. If α and β are variables, F_1 and F_2 formulae, and if $\ulcorner\alpha(F_1) = \beta(F_2)\urcorner$ contains no free variables, it is a sentence. It is true if and only if, for every numeral N, the result $F_1(N)$ of replacing all free occurrences of α in F_1 by N, and the result $F_2(N)$ of replacing all free occurrences of β in F_2 by N, are equivalent in $\boldsymbol{S}_A$ [i.e., are either both true in $\boldsymbol{S}_A$ or neither one true].
9. For any predicate $\ulcorner\alpha(F)\urcorner$ and numeral N, the expression $\ulcorner\alpha(F)N\urcorner$ is a sentence (as well as a formula) and is true if and only if the result F(N) of replacing all free occurrences of α in F by N, is true.
10. If F_1 and F_2 are formulae, so is $\ulcorner(F_1)\downarrow(F_2)\urcorner$. The free occurrences of any variable α are those of F_1 and those of F_2. If F_1 and F_2 are sentences then $\ulcorner(F_1)\downarrow(F_2)\urcorner$ is a sentence and is true if and only if neither F_1 nor F_2 is true.[10]

Note: The notation 'F(N)' of (8) or (9) will also be used for an arbitrary term t, not necessarily a numeral—i.e., $\ulcorner F(t)\urcorner \underset{df}{=}$ the result of substituting freely the term t for the free variable of F.

Gödel Numbering. For any expression E, let $\sigma(E)$ be the string of Arabic numerals obtained by replacing S_1 by the Arabic numeral '1', S_2 by '2', ..

[10] We could have used the single primitive '$\subset$' (class inclusion) in place of the joint use of '$\downarrow$' and '=' (as occurring between abstracts). We would then have a system formulated in a logic based on inclusion and abstraction (in the sense of Quine). All results of this paper would still go through.

S_9 by '9'. This string $\sigma(E)$ designates (in decimal notation) a number, which we will call $g_0(E)$. We shall take for our Gödel number of E (written '$g(E)$' or '$\mathring{E}$') the number $g_0(E) + 1$.

It will facilitate our exposition if we identify the numbers with the numerals (strings of '1's) which designate them in S_A. Then we define the norm of E to be E followed by its own Gödel number.

Arithmetization of the Norm Function. The following extremely simple definition accomplishes all the arithmetization of syntax which we need:

Def. 1. $n(x) \underset{df}{=} x \cdot 10^x$.

Explanation. If x is the g.n. (Gödel number) of E, then $n(x)$ is the g.n. of the norm of E. Thus, for example, 37 is the g.n. of $S_3 S_6$. The norm of $S_3 S_6$ is $S_3 S_6 \underbrace{S_9 S_9 \ldots\ldots S_9}_{37}$, and its g.n. is

$$3\underbrace{699\ldots.9}_{37} + 1 = 37\underbrace{00\ldots0}_{37} = 37 \times 10^{37}.\text{[11]}$$

Semantical Normality. We say that an expression E satisfies the predicate H (relative to the Gödel correspondence g) if $H\mathring{E}$ is true. The set W of all expressions satisfying H is precisely the set defined by H, in the sense of Section 3. We also say that E satisfies the *formula* F (when F contains one free variable) if $F(\mathring{E})$ is true, and we shall also refer to the set W of all E which satisfy F as the set defined by the formula F. Now, the crucial role played by the class abstractors of S_A is that definability by a predicate, and definability by a formula, are thereby equivalent. This is an immediate consequence of Rule 9 of S_A since, if H is the abstraction of F, then E satisfies H $\Leftrightarrow$ $H\mathring{E}$ is true $\Leftrightarrow$ $F(\mathring{E})$ is true (by Rule 9) $\Leftrightarrow$ E satisfies F. Thus the sets respectively defined by H and F are the same.

A formula F_N will be called a *normalizer* of formula F if F_N is satisfied by just those expressions E whose norm satisfies F. In the light of the preceding paragraph, the statement that S_A is semantically normal is equivalent to the statement that every formula F (with one free variable) has a normalizer F_N (since F_N defines $\eta(W)$, when F defines W).

THEOREM 4. *S_A is semantically normal, relative to g.*

PROOF. We must show that every F has a normalizer F_N. Well, take F_N to the result of replacing the free variable α of F by $\alpha \cdot 10^\alpha$ [or rather, by the abbreviated form $\ulcorner(\alpha)\cdot((1111111111 \urcorner(\alpha))\urcorner$.]

11 Had we used g_0, rather than g for our Gödel correspondence, then, if x were the g.n. of E, the g.n. of the norm of E would have been $(x + 1)10^x - 1$, rather than $x \cdot 10^x$.

Then, for any number x, $F_N(x)$ and $F(n(x))$ have the same truth-values. Thus, for any expression E,

$$\text{E satisfies } F_N \Leftrightarrow F_N(\mathring{E}) \text{ is true}$$

$$\Leftrightarrow F(n(\mathring{E})) \text{ is true}$$

$\Leftrightarrow$ the norm of E satisfies F (since $n(\mathring{E})$ is the g.n. o norm of E!). Hence F_N is satisfied by those E whose norm satisfies F and i thus a normalizer of F.

COROLLARY 4.2. (*1*) *For every definable set of* $\boldsymbol{S}_A$, *there is a Tarski sentenc* (*2*) *The complement of the truth set* T *of* $\boldsymbol{S}_A$ *is not definable in* $\boldsymbol{S}_A$ (*relative to g* (*3*) T *itself is not definable in* $\boldsymbol{S}_A$ (*relative to g*). (*4*) *Any proposed axiomat zation of* $\boldsymbol{S}_A$ *such that the set of its theorems is definable in* $\boldsymbol{S}_A$ (*relative to g*) *semantically incomplete or inconsistent.*

(1) and (2) immediately follow from the preceding theorem, together wit the results of Section 3. In particular, in (1), to construct a Tarski sentenc for a set *W* defined by formula F, we first construct the normalizer F_N of by the method of the preceding theorem, then take the abstraction H of F_N and then follow H by its own Gödel number. Thus the Tarski sentence fc *W* is the norm of the abstraction of the normalizer of the formula whic defines *W*. (3) and (4) follow, since $\boldsymbol{S}_A$ contains negation (definable fro '↓').

REMARK. (4) of Corollary 4.2 can be thought of as one form of Gödel theorem. Definability in $\boldsymbol{S}_A$ is actually equivalent to definability in prot syntax (in the sense of Quine). Thus any formal system for $\boldsymbol{S}_A$ whose set theorems is protosyntactically definable will be semantically incomplete inconsistent. This is essentially similar to Quine's result that protosynt itself is not protosyntactically completable.

4.3. We have just shown a method for constructing normalizers whi works for the particular Gödel correspondence *g*, which we employe Actually, it will work for any Gödel correspondence relative to which t norm function (i.e., the function which assigns to each expression its nor is *strictly* definable, in the following sense:

A function f (from expressions to expressions) will be said to be *stric* defined by the term t (with one variable α), if for any two expressions E_1 a E_2, $E_1 = f(E_2)$ if and only if the numeral $\mathring{E}_1$ and the term $t(\mathring{E}_2)$ (viz., t result of substituting $\mathring{E}_2$ for α in t) designate the same number. This noti of strict definability is quite different from the usual much weaker notion definability of f, viz., the existence of a formula M with two free variab such that, for any E_1 and E_2, $E_1 = f(E_2)$ if and only if $M(\mathring{E}_1, \mathring{E}_2)$ is true. V can, in an obvious manner, extend both notions of definability to functions of more than one argument.

If now the norm function is *strictly* definable relative to g, then to construct a normalizer for F we simply replace all free occurrences of the free variable α of F by the term $\mathrm{t}(\alpha)$ which defines the norm function. Since this process nowhere makes use of quantifiers (or other identity of class abstracts),[12] or logical connectives, or more than one variable, then if we completely stripped $\boldsymbol{S}_A$ of its logical connectives, quantifiers, and all variables but one, the resulting vastly weaker system $\boldsymbol{S}_a$ would still be normal and, moreover, so would any extension of $\boldsymbol{S}_a$. Let us state this more precisely:

By the system $\boldsymbol{S}_a$, we mean the system whose signs are those of $\boldsymbol{S}_A$, except for '↓' and ''', and whose rules are those of $\boldsymbol{S}_A$, with the omission of Rules (7) and (10), and with Rule (2) changed to (2′), 'x' is a variable. By an extension of $\boldsymbol{S}_a$, we mean a system constructed from $\boldsymbol{S}_a$ by possibly adding additional signs and rules. Then the following theorem is a considerable strengthening of Theorem 4.1:

THEOREM 4.4. *Any extension* $\boldsymbol{S}_a'$ *of* $\boldsymbol{S}_a$ *is normal relative to any Gödel correspondence g, relative to which the norm function is strictly definable providing that whenever two terms* t_1 *and* t_2 *have the same designata,* $\mathrm{F}(\mathrm{t}_1)$ *and* $\mathrm{F}(\mathrm{t}_2)$ *have the same truth-values.*

4.5. NORMALITY OF $\boldsymbol{S}_A$ RELATIVE TO OTHER GÖDEL CORRESPONDENCES. If the norm function is definable (relative to g) in only the weaker sense, rather than strictly definable, then, although the above method of constructing normalizers is no longer available to us, we still have another method which will work for $\boldsymbol{S}_A$, but *not* for $\boldsymbol{S}_a$ (or any arbitrary extension thereof), since the construction depends on quantification.

Letting $\mathrm{N}(\alpha,\beta)$ be the formula which defines the norm function, we let $\mathrm{F_N} \underset{\mathrm{df}}{=} \ulcorner(\exists\beta)(\mathrm{N}(\beta,\alpha)\ \&\ \mathrm{F}(\beta))\urcorner$, where the existential quantifier is defined from the universal quantifier in the usual manner, the latter defined as previously indicated, and '&' is defined from '↓' in the usual manner. Then $\mathrm{F_N}$ is a normalizer of F. Hence,

THEOREM 4.6. *If the norm function is definable in* $\boldsymbol{S}_A$ *relative to a Gödel correspondence g, then* $\boldsymbol{S}_A$ *is normal relative to g.*

We lastly observe that if there is a formula $C(\alpha,\beta,\gamma)$ such that, for any expressions E_1, E_2, E_3, $\mathrm{E}_3 = \mathrm{E}_1\,\mathrm{E}_2$ if and only if $C(\mathring{\mathrm{E}}_1,\mathring{\mathrm{E}}_2,\mathring{\mathrm{E}}_3)$ is true (which

12 As indicated at the beginning of Section 4, quantification is defined, using a formula which employs the identity sign between class abstracts.

we express by saying that concatenation is definable, relative to g) and if there is a formula $G(\alpha,\beta)$ such that, for any expressions E_1 and E_2, E_1 is the Gödel numeral of E_2 if and only if $G(\mathring{E}_1,\mathring{E}_2)$ is true (which we express by saying that g itself is definable relative to g), then the formula $\ulcorner(\exists\gamma)(G(\gamma,\beta).\&.C(\beta,\gamma,\alpha))\urcorner$ defines the norm function and $\boldsymbol{S_A}$ is normal. Hence

THEOREM 4.7. *A sufficient condition for* $\boldsymbol{S_A}$ *to be normal, relative to g, is that concatenation and g itself both be definable, relative to g.*

REMARK. Gödel correspondences satisfying the hypothesis of Theorem 4.7 include all those that are *effective* (i.e., include all those g such that the function h, which assigns to each number x the Gödel number of (the numeral designating) x, is a recursive function.[13] This, in conjunction with previous results, yields the proposition that, relative to any effective Gödel correspondence g, the truth set of $\boldsymbol{S_A}$ is not definable. This, in essence, is Tarski's Theorem.

5. *Concluding Remarks: Diagonalization vs. Normalization.* We should like, in conclusion, to compare the norm function, used throughout this paper, with the more familiar diagonal function, used for systems in standard formalization.

Firstly, to sketch a general account of diagonalization,[5] analogous to Section 3 for normalization, we consider now an arbitrary language $\boldsymbol{L}$ which (like $\boldsymbol{S}$ of Section 2) contains expressions, sentences, true sentences, and individual constants. Instead of predicates, however, we now have certain expressions called 'formulas' and others called 'variables,' and certain occurrences of variables in formulas termed 'free occurrences,' subject to the condition that the substitution of individual constants for all free occurrences of variables in a formula always yields a sentence. We again have a Gödel correspondence mapping each expression E onto an individual constant $\mathring{E}$. For any formula F with one free variable α and any expression E, we define F(E) as the result of substituting $\mathring{E}$ for all free occurrences of α in F. The expression F(F) is defined to be the *diagonalization* of F. The set

[13] e.g., the correspondence g_0. To show that, relative to g_0, the norm function is weakly definable, we must construct a formula $\ulcorner\phi(\alpha,\beta)\urcorner$ such that, for any two numbers n and m, $\phi(n,m)$ is true $\Leftrightarrow m+1=(n+1).10^n$ (cf. (9)). We first define addition as follows: Add $(\alpha,\beta,\gamma) \underset{\text{df}}{=} n^{\alpha}.n^{\beta}=n^{\gamma}$ (where we take n any number $\neq 1$). Then we define $\phi(\alpha,\beta) \underset{\text{df}}{=} (\exists\gamma)$ [Add $(\alpha,1,\gamma)$ & Add $(\beta,1,\gamma.10^{\alpha})$] (this construction can be simplified by introducing descriptors). Thus the tricky correspondence g_0+1, which we used, was introduced only for purposes of simplicity, and is certainly not necessary for the success of our programme.

f all E such that F(E) is true, is called the set *defined* by F. For any set W, 'e define $D(W)$ as the set of all F whose diagonalization is in W. Then the nalogue of Theorem 3.1 is 'A sufficient condition for the existence of a 'arski sentence for W is that $D(W)$ be definable.' Hence also, $D(\bar{W})$ is not efinable. We would then define normality, for such a language L, by the ondition that whenever W is definable in L, so is $D(W)$. Then all other ieorems in Section 3 have their obvious analogues.[14]

To apply these general notions to systems in standard formalization, e.g., lementary arithmetic, we would have, in analogy with the notion 'nor-ıalizer', that of 'diagonalizer', where a diagonalizer F_D of a formula F vould be a formula satisfied by just those expressions whose diagonalization atisfied F. Then, if W is defined by F, and if there exists a diagonalizer F_D or F, then the diagonalization of F_D (which is $F_D(F_D)$), is the Tarski entence for W.

This is essentially the classical construction. The construction of the iagonalizer F_D is considerably more involved than the construction of the ormalizer F_N. Again, we might say, this is due to the fact that concate-ation is easier to arithmetize than substitution.

14 We can profitably avoid repetition of analogous arguments for S and L by egarding both as special cases of a more general structure. This approach will be resented in a forthcoming paper, 'Abstract Structure of Unsaturated Theories', in vhich we study, in considerable generality, the deeper properties of undecidable ystems, uncovered by Gödel and Rosser.

V

INFORMAL RIGOUR AND COMPLETENESS PROOFS

Georg Kreisel

It is a commonplace that formal rigour consists in setting out formal rules and checking that a given derivation follows these rules; one of the more important achievements of mathematical logic is Turing's analysis of what a formal rule is. Formal rigour does *not* apply to the discovery or choice of formal rules nor of notions; neither of basic notions such as *set* in so-called classical mathematics, nor of technical notions such as *group* or *tensor product* (technical, because formulated in terms of an already existing basic framework).

The 'old fashioned' idea is that one obtains rules and definitions by analysing intuitive notions and putting down their properties. This is certainly what mathematicians thought they were doing when defining length or area or, for that matter, logicians when finding rules of inference or axioms (properties) of mathematical structures such as the continuum. The general idea applies equally to the so-called realist conception of mathematics which supposes that these intuitive notions are related to the external world more or less as the number 4 enters into configurations consisting of 4 elements, and to the idealist conception which denies this or, at least considers this relation as inessential to mathematics. What the 'old fashioned' idea assumes is quite simply that the intuitive notions are *significant*, be it in the external world or in thought (and a *precise* formulation of what is significant in a subject is the result, not a starting point of research into that subject).

Informal rigour wants (i) to make this analysis as precise as possible (with the means available), in particular to eliminate doubtful properties of the intuitive notions when drawing conclusions about them; and (ii) to extend this analysis; in particular not to leave undecided questions which can be decided by full use of evident properties of these intuitive notions. Below the principal emphasis is on intuitive notions which do not occur in ordinary mathematical practice (so-called new *primitive notions*), but lead to new

From *Problems in the Philosophy of Mathematics*, ed. Imre Lakatos (North-Holland Publishing Company, Amsterdam, 1967), pp. 138–57. Reprinted, with a postscript, by permission of the publishers and the author. [The last 14 pages of this essay, pp. 158–71 are omitted here, together with some sentences referring to them. Ed.]

ioms for current notions. We give three applications, mostly following e 'old fashioned' idea of pushing a bit farther than before the analysis of e intuitive notions considered. Section 1 concerns the difference between miliar independence results, e.g. of the axiom of parallels from the other ioms of geometry, on the one hand and the independence of the con- uum hypothesis on the other; the difference is formulated in terms of gher-order consequence. Section 2 deals with the relation between in- itive logical consequence on the one hand and so-called semantic resp. ntactic consequence on the other. Section 3 [is omitted here—Ed.]. ctions 2 and 3 affect completeness questions for classical and intuitionist edicate logic, which accounts for the title of this talk; quite generally, oblems of completeness (of rules) involve informal rigour, at least when ne is trying to decide completeness with respect to an intuitive notion of nsequence.

(0) *The case against informal rigour* (or: antiphilosophic doctrines). The esent conference showed beyond a shadow of doubt that several recent sults in logic, particularly the independence results for set theory, have ft logicians bewildered about what to do next: in other words, these results not 'speak for themselves' (to these logicians). I believe the reasons derlying their reaction, necessarily also make them suspicious of informal gour. I shall try to analyse these reasons here.

(a) *Doctrinaire objections* (*pragmatism, positivism*). Two familiar objec- ons to informal rigour are these:

(i) Why should one pay so much attention to intuitive notions? What we ant are definitions and rules that are *fruitful*; they don't have to be *faithful* notions that we have already. One might perhaps add: these notions are rmed without highly developed experience; so why should they be ex- cted to be fruitful?

Besides this (pragmatic) objection we have a more theoretical (positivist) jection.

(ii) These intuitive notions, in particular the (abstract) notions of validity, t, natural number or, so as not to leave out intuitionism, intuitively con- ncing proof, are illusions. When one examines them one finds that their lid content lies in what we do, in how we act; and, in mathematics, this is ntained in the formal operations we perform.

A certain superficial plausibility cannot be denied to these objections. irst, when some abstract intuitive notion turns out to be equivalent, at ast in a certain context, to a positivistic relation, i.e., one definable in rticularly restricted terms, this has always important consequences. For stance (for detail, see Section 2) logical consequence applied to first order formulae, is equivalent to formal derivability; and first order axiom systems permit a more general theory than higher order systems. Consequently, at a particular stage, the (pragmatically) most rewarding work in the subject may consist quite simply in exploiting the discovery of such an equivalence Second, one may be impressed by the slow progress of work on some of the intuitive notions, particularly those associated with traditional philosophic questions: pragmatism discourages such work, and positivism tries to give theoretical reasons for the slow progress. Now, objectively, such a negative attitude is not supported by the facts because progress was also slow in case where decisions *were* eventually obtained. (About 30 years between Hilbert's first formulation of his finitist programme, cf. (c), p. 84, and Gödel' incompleteness theorems; nearly a further 30 years till a precise analysis o finitist proof was attempted.) But, subjectively, if a particular person i discouraged by the slow progress he had surely better find himself anothe occupation. Certainly, scientifically speaking, one is in a wholly futile posi tion if one finds oneself stuck both with philosophy as a profession and with antiphilosophical views such as pragmatism or positivism (perhaps, afte having been attracted by traditional questions in one's youth). For, havin repudiated specifically philosophical notions one is left with those that ar also familiar to specialists in other fields: what jobs can one hope to do a well as these specialists? including the jobs of clarification or explicatio (if they are to be done in current terms)? I think this futility is felt quit consciously by many of the people involved.

Having granted all this: what is wrong with (i) and (ii)? Quite simply this *Though they raise perfectly legitimate doubts or possibilities, they just do no respect the facts, at least the facts of actual intellectual experience.* This i particularly irritating because pragmatism pays so much lip service to ex perience, and positivism claims to be empirically minded.

Ad (i). Let us even take for granted that we know roughly what is fruitfu after all, here again (i.e. as in the case of significance) a precise formulatio may only be possible after a good deal of experience. Perhaps we do no know a general reason why intuitive notions *should* have stood the tests o experience well; as one sometimes says: they might not have done so. Bu the fact remains that they, or, at least, many of them, have. Reflection show that we certainly couldn't have what we understand by 'science' if the hadn't. Instead of trying to find reasons for, or limitations of, this supe ficially remarkable situation, (i) denies its existence! (a most unpragmat pragmatism). Two related so-called pragmatic principles are to be men tioned. One says that one must treat each problem 'on its merits'; one migh have to; but taken literally this would leave little room for general theory o r the distinction between what is fundamental and what is secondary. Or gain (in mathematics), one sometimes criticizes complacently 'old shioned' disputes on the right definition of measure or the right topology, cause there are several definitions. The most striking fact here is how *few* em to be useful: these haven't dropped from heaven; they, obviously, ere formulated *before* their applications were made, and they were not nerally obtained by trial and error. If they had been so obtained, mathe-aticians shouldn't be as contemptuous as they are about the study of little riants in definitiôns. Similar remarks apply to the choice of axioms; but nce this is of direct logical importance the subject will be taken up in its oper place in (b).

Ad (ii). Clearly, if, consciously or unconsciously, one insists on analysing e 'solid content' in positivistic, in particular, formal, terms this is what one ill find. Though more specific points about formalism and formalization e taken up throughout this paper in particular (c) and Section 2 below, me matters of principle are in order here. It might have turned out that e notions which present the most serious difficulties in practice are indeed stract ones. But, quite naïvely, this is not so: knowing whether two scriptions *mean* the same is often no harder than knowing if they *read* the me! Equally, as was mentioned on p. 79 above, sometimes it does turn it that some notions are fully represented in formalistic terms: but this has be verified and Section 1 shows limitations. Perhaps one should dis-nguish between formalism (and positivism), which is merely a negative tiphilosophical doctrine, and a *mechanistic* conception of reasoning nechanism' in the sense of Turing), which would lead one to expect a full rmalistic analysis of actual reasoning. It is to be remarked that, so far, the ost that has ever been shown in support of this conception is that in rtain areas (e.g., elementary logic, Section 2) reasoning *could* be mech-ical in the sense that a mechanism would get the same results; not that it , i.e., that it would follow the same routes. It may be that the mechanistic nception is the only moderately clear idea of reasoning that we have at esent. But a good positivist should not conclude from this that therefore is idea is correct.

(b) *Unreliability of some intuitive notions; the role of formalization in their alysis.* A much more serious point than the portmanteau objections (i) d (ii) under (a) concerns specific abstract notions, for instance—to take e most famous example—the notion of set. *Have not the paradoxes shown e complete unreliability of our intuitive convictions at least about this par-ular notion?*

First of all, historically speaking, this couldn't be farther from the truth!

Wasn't Cantor a misunderstood martyr in the face of widespread re actionary prejudice against employing the notion of set (or, as it was the called: class) in mathematics? If so, the paradoxes supported the intuitiv convictions of those reactionaries.

It is probably true to say that the reactionary caution was due to this *class* presented itself as a vague notion, or, specifically, a mixture of notion including (i) finite sets of individuals (i.e. objects without members), or (ii sets *of* something (as in mathematics, sets of numbers, sets of points), bu also (iii) properties or *intensions* where one has no *a priori* bound on th extension (which are very common in ordinary thought but not in mathe matics). If we are thinking of sets *of* something, e.g., of objects belonging t a, then the comprehension axiom is to be restricted to read (for any pro perty P)

$$\exists x\, \forall y(y \in x \leftrightarrow [y \in a \;\&\; P(y)]);$$

but if we are thinking of properties, given in intension, whose range c definition is not determined, we may well have (with variables ranging ove properties)

$$\exists Q\, \forall R[Q(R) \leftrightarrow P(R)]:$$

only one had better remember that these properties are not everywhe defined and so the laws of two valued logic are not valid. (So, to be precis the logical symbols have different meaning in the two cases.)

Now, the reactionaries were wrong because at least one element of th mixture (namely: set *of* something), first described clearly by Russell an especially, Zermelo, has proved to be marvellously clear and comprehensiv But before this analysis the prospects were not rosy: what one could ho to do was to put down *assertions which are satisfied by all elements of th mixture*, and it just didn't look as if this 'common part' was going to lead t a mathematically rich theory; more precisely, to something in terms c which anything like (then) current mathematics could be interpreted. I contrast, axioms which are evidently valid for the particular notion isolate by Zermelo (cumulative type structure) give a formal foundation (even) fc a great deal of present day mathematical practice.

The main problem which, in my opinion, the paradoxes present can b put quite well in old fashioned language: *what are the proper laws* (the 'logic *satisfied by the intensional element of the crude mixture*, in particular, th element (iii) which satisfies the unrestricted comprehension axiom?[1]

1 The recursion theorem for partial (recursive) functions is analogous to such comprehension axiom: this might serve as a model 'in the small' for *very* abstra notions which again satisfy a comprehension axiom without type restriction. It shou

Two *conclusions* about matters discussed at this meeting follow. First, Zermelo's analysis furnishes an instance of a rigorous *discovery of axioms* (for the notion of set). To avoid trivial misunderstanding note this: What one means here is that the intuitive notion of the cumulative type structure provides a coherent *source* of axioms; our understanding is sufficient to avoid an endless string of ambiguities to be resolved by further basic distinctions, like the distinction above between abstract properties and sets *of* something. Pragmatically speaking, cf. (a) above, one does not have to put up with an *ad hoc* collection of different axioms for different 'purposes' (though *ad hoc* considerations may be needed to show which of the new axioms are relevant to a particular purpose). Denying the (alleged) *bifurcation* or *multifurcation of our notion of set of the cumulative hierarchy* is nothing else but asserting the properties of our intuitive conception of the cumulative type-structure mentioned above. This does not deny the established fact that, in addition to this basic structure, there are also technically interesting non-standard models. From the present point of view, the importance of adding strong *axioms of infinity* (existence of large ordinals α) is clear. For, adding them does not restrict the unconditional assertions we can make about initial segments of the hierarchy below α; but leaving them out stops us from saying anything unconditional beyond α. For instance, consider the axiom of infinity (existence of ω, which is included in Zermelo's axioms); an arithmetic theorem A proved in Zermelo's theory, an arithmetic theorem B proved from the Riemann hypothesis only. Without the assumption of ω, both A and B are hypothetical: 'statistically', I suppose, 'equally' justified, since after all, the Riemann hypothesis has not led to a contradiction (otherwise it would be refuted). But, intuitively, A is established and B is not. Unless one denies the validity of this distinction, leaving out the assumption of ω conflicts with requirement (ii) on informal rigour on pp. 78 ff.

Second, the actual *formulation* of axioms played an auxiliary rather than basic role in Zermelo's work: the intuitive analysis of the crude mixture of notions, namely the description of the type structure, led to the good axioms: these constitute a record, not the instruments of clarification. And a similar conceptual analysis will be needed for solving the problem of the paradoxes.

(c) *Formalization.* What has been said above about the formulation of axioms, applies even more to the formulation of rules of inference (for

be mentioned that the views of the preceding two paragraphs contradict those (implicit) in the literature; e.g., Mostowski, in these Proceedings, sees the kernel of set theory in a kind of common part of different notions of set, and Rasiowa and Sikorski regard the paradoxes as a dead (fruitless) issue. (H. Rasiowa and R. Sikorski, *The Mathematics of Metamathematics*, Warsaw (1963)).

further details, see Section 2). In fact, on quite *general* grounds, one would expect the role of formalization to be always auxiliary in the analysis of notions. After all, the job of formalization is to record and codify arguments without distinguishing the good from the bad. And, leaving generalities, if one considers the so-called 'crises' in mathematics, one never disagreed about the inferences themselves, but either about the axioms (comprehension axiom) or about the *rules* of inference (law of the excluded middle in the intuitionistic criticism). So, *precision of the notion of consequence was not a primary issue*. If one believes that precision is the principal objective of formalization, and formalization of logic, one cannot be surprised that French mathematicians used to think of logic as the hygiene of mathematics and English mathematicians of logic as a matter of dotting the *i*'s and crossing the *t*'s. (What logic does is to study notions which were previously not recognized at all, or, if recognized, only used heuristically, and not made an object of detailed study; among them traditional philosophic notions.)

It seems certain that, for the psychological process of understanding, formalization is indeed important; only I do not have a good analysis. But there are some quite clear reasons why the role of formalization for *fundamental* analysis should be overvalued.

First and foremost, there are probably quite a few who scorn Hilbert's programme, but hold on to formalization as a kind of collective reflex. Hilbert wanted to *show* (as the positivist in (a) above should have done) that one really lost nothing by confining oneself to formal operations, and he found a way of expressing this in the form of a mathematical problem, namely his programme. Axiomatization, in fact strict formalization of inference, was *essential* for the very formulation of this programme. Incidentally, there was no idea of *rejecting* on general grounds the notion of second order consequence (Section 2), but of showing its equivalence, at least in suitable contexts, to formal derivability, as had been done for first order logical consequence. Unless one had been convinced of the latter equivalence one would never have engaged in Hilbert's programme.

Second, explicit formulation of axioms and rules undoubtedly plays a big role in everyday work of logicians. *Examples*: (i) If the basic concepts are accepted, one can make deductions *from* axioms clearer by eliminating those formally unnecessary to the conclusion; but, e.g., in Zermelo's analysis, to make the meaning of the basic concepts clear, he made them *more* specific. (ii) Suppose one wants to explain why a certain question happens to be open; one guesses a formal system, i.e., properties of the intuitive notion, which mathematicians are likely to use; one supports this guess by showing that current mathematics follows from these axioms, and one explains the situation by showing that this question is not decided in the formal system

(I personally like this sort of thing). (iii) If one wants to remember a proof one remembers the axioms used in the proof: the less there are the less proofs have to be remembered.

Of course all this goes to show that a lot of the everyday work of logicians is not concerned with fundamental analysis at all (even if one would like it to be). As someone (nearly) said at Balaclava: *C'est magnifique, mais ce ne sont pas les fondements.*

1. *Higher-Order Axiomatizations and Independence Proofs.* This section takes the precise notion of set (in the sense of the cumulative[2] type structure of Zermelo, (b), p. 82) as starting point and *uses* it to formulate and refine some intuitive distinctions. Specifically, one analyses concepts according to the *order* of the language needed to define them. The connection between this technical exposition (in the sense of p. 78) and informal rigour appears in (c) below.

It is clear that the matter is technical because the very idea of a definition requires an interpretation of the language, for instance of the logical symbols. In the present case they are interpreted by means of the (set-theoretic) operations of complementation, union and projection. The languages used are those of *predicate logic* (for instance, in Church's book.)

More or less familiar examples. Such notions as *equivalence relation* or *order* are defined by first-order formulae A_E, A_O resp., i.e., in ordinary predicate calculus, in the sense that a structure consisting of a domain a and a relation b on a (i.e., $b \subset a \times a$) satisfies A_E, A_O if and only if b is an equivalence relation, resp. an order on a. We are nowadays very familiar with this but it was a not at all obvious discovery that intuitive notions of equivalence or order could be defined in terms of such a simple language as predicate logic of first order. Equally it was a discovery that the only concepts definable in *this* way which are unique up to isomorphism are finite structures. So, both the expressive power and the limitations of first-order language came as a surprise.

The familiar classical structures (natural numbers with the successor relation, the continuum with a denumerable dense base, etc.) are definable by *second*-order axioms, as shown by Dedekind. Zermelo showed that his

2 Professor Bar-Hillel asked in discussion if one starts with a structure containing *Urelemente*, i.e., objects which have no elements but are different from $\emptyset$. Evidently, the answer is yes, because most concepts do not present themselves as concepts of sets at all (apples, pears). But it is a *significant theorem* that the classical structures of mathematics occur already, up to isomorphism, in the cumulative hierarchy *without* individuals. For the reduction of mathematics to set theory it is important to convince oneself that intuitively significant features are invariant under isomorphism, or, at least, classes of isomorphisms definable in set-theoretic terms, e.g., recursive ones.

cumulative hierarchy up to ω or $\omega + \omega$, or $\omega + n$ (for fixed n) and other important ordinals is equally definable by second-order formulae. Whenever we have such a second-order definition there is associated a *schema* in first order form (in the language considered): For instance, in Peano's axiom

$$\forall P[\{P(0) \,\&\, \forall x[P(x) \rightarrow P(x+1)]\} \rightarrow \forall x P(x)]$$

one replaces the second-order quantifier P by a list of those P which are explicitly defined in ordinary first-order form (from + and ×, for instance). A moment's reflection shows that the evidence of the first-order axiom schema derives from the second-order schema: the difference is that when one puts down the first-order schema one is supposed to have convinced oneself that the specific formulae used (in particular, the logical operations) are well defined in any structure that one considers; this will be taken up in (b) below. (*Warning*. The choice of first-order schema is not uniquely determined by the second-order axioms! Thus Peano's own axioms mention explicitly only the constant 0 and the successor function S, not addition nor multiplication. The first-order schema built up from 0 and S is a weak, incidentally decidable, subsystem of classical first-order arithmetic above, and quite inadequate for formulating current informal arithmetic. Informal rigour requires a much more detailed justification for the choice of + and × than is usually supplied.)

An interesting example of a concept that needs a third-order definition is that of measurable cardinal.[3] Such concepts are rare; for, whenever theoretically a whole hierarchy presents itself, in practice one only uses the first few levels—or a notion outside that hierarchy altogether; two concepts may be mentioned here which are not definable by any formula in the whole hierarchy of languages of predicate logic.

Evidently, neither the cumulative type structure itself not the structure consisting of the ordinals with the ordering relation is definable by any formula of finite (or transfinite) order: for (at least, usually) one requires that the universe of the structures considered be a set, and no set is isomorphic to the totality of all ordinals, let alone of all sets.

NB. The (somewhat crude) classification in terms of *order* of the language considered has recently been refined by the use of *infinite* formulae, for instance, infinite first-order formulae. These or, at least, important classes $\mathscr{C}$ of them are *intermediate* between first-order and second-order formulae since any structure definable by a formula in $\mathscr{C}$ is also definable by means of a (finite) second-order formula, but not conversely. The ordinal ω is an example.

[3] W. Hanf and D. Scott, 'Classifying inaccessible cardinals', *Notices Amer. Math. Soc.* **8** (1961), p. 445.

It is clear that what is achieved in the case of ω is a technical analysis of ertain sets of integers by means of the notion of integer (which is used ssentially in the theory of infinite formulae)—perhaps, as one uses induction to define, and obtain results about, prime numbers. The notion of nteger itself is not analysed in this way.

What is much more interesting than this obvious remark is the fact that a ich theory of infinite formulae can be developed: specifically, many useful heorems about finite first-order formulae can be extended to infinite ones cunningly chosen) but not to finite or infinite second-order formulae.

(a) A *reduction* of assertions about higher order consequence to first-rder statements in the language of set theory. Since the notion of realization f a formula (or, of model) of any given order is formulated in terms of the asic notion of set one may expect that, e.g., $A \vdash_2 B$, i.e., B is a consequence f the second-order formula A, is expressed by a first-order formula of set heory. More precisely, expressed by such a formula when the quantifiers re interpreted to range over all sets of the cumulative type structure. One xpects this, simply because it is always claimed that this first-order language adequate for all mathematics; so if it weren't adequate for expressing econd-order consequence, somebody would have noticed it.[4]

A simple calculation verifies this; moreover, *the definition uses exactly the ame basic notions as that of first-order consequence*: only instead of a uantifier $\forall a$ (over all sets) followed by a formula whose quantifiers are estricted to a, i.e. $(\forall b \in a)$, one has a formula containing also quantifiers $\forall b \subset a)$.

As a corollary, any conclusion that we may formulate in terms of second-rder consequence can also be formulated by means of a first-order assertion bout the cumulative type structure. However, *heuristically*, that is for nding the first-order assertion, it may be very useful to think in terms of econd-order assertions.

Example. Let $\mathscr{Z}$ be Zermelo's axiom with the axiom of infinity, and let H be the (canonical) formulation of the continuum hypothesis in the llowing form: if C_ω is the collection of hereditarily finite sets without dividuals, $C_{\omega+1} = C_\omega \cup \mathfrak{P}(C_\omega)$, $C_{\omega+2} = C_{\omega+1} \cup \mathfrak{P}(C_{\omega+1})$, CH states that

$$X \subset C_{\omega+1} \to (\overline{\overline{X}} \leqslant \overline{\overline{C}}_\omega \vee \overline{\overline{X}} = \overline{\overline{C}}_{\omega+1}),$$

[4] One cannot be 100 per cent sure: for instance, consider the so-called truth definin. We have here a set T of natural numbers, namely Gödel numbers $\bar{\alpha}_i$ of first-order rmulae of set theory, such that $n \in T \leftrightarrow \exists i(n = \bar{\alpha}_i \;\&\; \alpha_i)$, i.e., T is defined by

$$(n = \bar{\alpha}_1 \;\&\; \alpha_1) \vee (n = \bar{\alpha}_2 \;\&\; \alpha_2) \vee \ldots$$

Tarski emphasized, T is not definable by means of a first-order formula (in the ecise sense above).

which is expressed by means of quantifiers over $C_{\omega+2}$. As Zermelo pointe out (see above), if we use the current set-theoretic definition $Z(x)$ of th cumulative hierarchy, in any model of $\mathscr{Z}$, this formula Z defines a C_σ for limit ordinal $\sigma > \omega$. Consequently we have

$$(\mathscr{Z} \vdash_2 CH) \vee (\mathscr{Z} \vdash_2 \sim CH).$$

Note that *CH* is formulated in first-order language of set theory.

(b) *Distinctions* formulated in terms of higher-order consequence. I contrast to the example on *CH* above, Fraenkel's replacement axiom is no decided by Zermelo's axioms (because $\mathscr{Z}$ is satisfied by $C_{\omega+\omega}$, and Fraenkel' axiom not); in particular it is independent of Zermelo's second-orde axioms while by Cohen's proof, *CH* is only independent of the *first-orde schema* (associated with the axioms) of Zermelo–Fraenkel.

This shows, first of all, the (mathematical) fact that *the distinction betwee second-order consequence and first-order consequence* (from the schema) *not trivial.*

Secondly, it shows a *difference* between the independence of the axiom c parallels in geometry on the one hand and of *CH* in first-order set theory In geometry (as formulated by Pasch or Hilbert) we also have a *second-orde* axiom, namely the axiom of continuity or Dedekind's section: *the parall axiom is not even a second-order consequence of this axiom*, i.e., it corre sponds to Fraenkel's axiom, not to *CH*.

Finally, consider the empirical fact that nobody was astonished by th independence of Fraenkel's axiom, but many people were surprised b Cohen's result. This reaction is quite consistent with my assertion above th the evidence of the first-order schema derives from the second-order axion Even if one explained to a mathematician the distinction above he woul marvel at the ingenuity required to *exploit* it; in his own work he never giv a second thought to the form of the predicate used in the comprehensic axiom! (This is the reason why, e.g., Bourbaki is extremely careful to isola the assumptions of a mathematical theorem, but never the axioms of s theory implicit in a particular deduction, e.g., what instances of the com prehension axiom are used. This practice is quite consistent with th assumption that what one has in mind when following Bourbaki's proofs the second-order axiom, and the practice would be horribly unscientific one really took the restricted schema as basic.)

(c). Connection between *informal rigour and the notion of higher-ord consequence*. The first point to notice is that his notion is needed for the ve formulation of the distinction above. This illustrates the weakness of th positivist doctrine (ii) in (a), p. 79, which refuses to accept a distinctic unless it is formulated in certain restricted terms. (NB. Of course *if* o ants to study the formalist reduction, Hilbert's program of (c), p. 84, the striction is not only acceptable, but necessary. But the fact that the in-itively significant distinction above cannot be so formulated, reduces the undational importance of a formalistic analysis, by requirement (ii) of formal rigour.)

Next it is not surprising that there is a certain *asymmetry* between the role higher-order consequence for derivability results (Section 1(a)) and its le in independence results (Section 1(b)). The same is familiar, e.g., from cursion theory. Thus to establish negative (i.e., unsolvability) results one ll aim in the first place to show recursive unsolvability, while to show lvability one gives a *particular* schema and a *proof* showing that the hema works. (Similarly, cf. end of the introduction; even if a problem *is* cursively solvable, one may wish to explain why it has not been solved: by owing, e.g., that there is no schema of a given kind which can be *proved* work by given methods, or else by showing that calculations are too long.) iriously enough, this obvious point is sometimes overlooked.

Finally, and this is of course the most direct link between the present and e main theme of this article, second-order decidability of *CH* (in the ample of (a) above) suggests this: new primitive notions, e.g. properties natural numbers, which are *not* definable in the language of set theory ich as in the footnote[4] on p. 87), may have to be taken seriously to decide *I*; for, what is left out when one replaces the second-order axiom by the iema, are precisely the properties which are not so definable. But I am re I don't know: the idea is totally obvious; most people in the field are so customed to working with the restricted language that they may simply t succeed in taking other properties seriously; and, finally, compared th specific examples that come to mind, e.g. the footnote on p. 87, the called axioms of infinity[5] which are formulated in first-order form are re efficient.

2. *Intuitive Logical Validity, Truth in all Set-theoretic Structures, and rmal Derivability*. We shall consider formulae α of finite order (α^i denot- formulae of order i), the predicate $Val\alpha$ to mean: α is intuitively valid, : α is valid in all set-theoretic structures, and $D\alpha$: α is formally derivable means of some fixed (accepted) set of formal rules.

For reference below. $V\alpha$ is definable in the language of set theory, and for ursive rules $D\alpha$ is definable *uniformly*, i.e. for each $\sigma \geqslant \omega$, the same mula defines D when the variables range over C_σ. Below, we shall also isider $V_C\alpha$: validity in classes (i.e. the universe of the structure is a class

K. Gödel, 'Remarks before the Princeton Bicentennial Conference on problems nathematics', *The Undecidable* (ed. M. Davis), New York (1965), pp. 84–8.

and the relations are also classes) at least for formulae of first and secon order.

What is the Relation Between Val and V?

(a) *Meaning of Val.* The intuitive meaning of *Val* differs from that of in one particular: $V\alpha$ (merely) asserts that α is true in all structures in th cumulative hierarchy, i.e., in all sets in the precise sense of *set* above, whil $Val\alpha$ asserts that α is true in *all* structures (for an obvious example of th difference, see pp. 90–91). A current view is that the notion of arbitrar structure and hence of intuitive logical validity is so vague that it is absur to ask for a proof relating it to a precise notion such as V or D, and that th most one can do is to give a kind of plausibility argument.

Let us go back to the fact (which is not in doubt) that one reasons i mathematical practice, using the notion of consequence or of logical co sequence, freely and surely (and, recall p. 84, the 'crises' in the past i classical mathematics by (c), p. 83, were not due to lack of precision in th notion of consequence). Also, it is generally agreed that at the time of Freg who formulated rules for first-order logic, Bolzano's set-theoretic definitio of consequence had been forgotten (and had to be rediscovered by Tarski yet one recognized the validity of Frege's rules (D_F). This means th implicitly

$$\forall i\, \forall \alpha(D_F\, \alpha^i \rightarrow Val\alpha^i)$$

was accepted, and therefore certainly *Val* was accepted as meaningful.

Next, consider the two alternatives to *Val*. First (e.g. Bourbaki) 'ult mately' inference is nothing else but following formal rules, in oth words D is primary (though now D must not be regarded as defined se theoretically, but combinatorially). This is a specially peculiar idea, becau 99 per cent of the readers, and 90 per cent of the writers of Bourbaki, do have the rules in their heads at all! Nobody would expect a mathematici to work on groups if he did not know the definition of a group. (By Sectio 1(b), the notion of set is treated in Bourbaki like *Val*.)

Second, consider the suggestion that 'ultimately' inference is semantic i.e. V is meant. This too is hardly convincing. Consider a formula α^1 with t binary relation symbol E as single non-logical constant; let α_ϵ mean that is true when the quantifiers in α range over all sets and E is replaced by t membership relation. (Note that α_ϵ is a first-order formula of set theor Then intuitively one concludes:

If α is logically valid then α_ϵ, i.e. (in symbols): $Val\alpha \rightarrow \alpha_\epsilon$. But o certainly does not conclude *immediately*: $V\alpha \rightarrow \alpha_\epsilon$; for α_ϵ requires that be true in the structure consisting of all sets (with the membership relatio its universe is not a set at all. So $V\alpha$ (α is true in each set-theoretic structu oes not allow us to conclude α_ϵ 'immediately': this is made precise by eans of the results in (b) and (c) below. On the other hand one does accept

$$\forall i\, \forall \alpha(Val\alpha^i \to V\alpha^i)$$

e moment one takes it for granted that logic *applies* to mathematical ructures.

Nobody will deny that one knows more about Val after one has established s relations with V and D; but that doesn't mean that Val was vague before.

In fact we have the *theorem*:

For $i = 1$, given the two accepted properties of *Val* above,

$$\forall \alpha^1(Val\alpha \leftrightarrow V\alpha) \quad \text{and} \quad \forall \alpha^1(Val\alpha \leftrightarrow D\alpha).$$

The proof uses Gödel's completeness theorem: $\forall \alpha^1(V\alpha \to D\alpha)$. Comned with $\forall \alpha^1(D\alpha \to Val\alpha)$ above, we have $\forall \alpha^1(V\alpha \leftrightarrow D\alpha)$, and with $i\, \forall \alpha(Val\alpha^i \to V\alpha^i)$ above: $\forall \alpha^1(V\alpha \leftrightarrow Val\alpha)$. Without Gödel's completeess theorem we have from the two accepted properties of *Val*: $\alpha^1(D\alpha \to V\alpha)$, incidentally a theorem which does not involve the primitive otion *Val* at all.

At least, *Val* is not too vague to permit a *proof* of its equivalence to V for rst-order α, by use of the properties of *Val* above!

(b) *The relation between* $V\alpha^1$ (α containing a binary E as single non logical onstant) *and* α_ϵ. To discuss this it is convenient to use the theory of exicitly definable properties (usually called: theory of classes) and the lation

$$Sat(A, B; \alpha)$$

mean: the property A and the relation B ($\subset A \times A$) satisfy α. We can present finite sequences of classes $A_1, \ldots, A_p$ by a single class $= \{\langle n,x\rangle : n \leqslant p \;\&\; x \in A_n\}$. If each A_i is explicitly definable so is A. Now, standard techniques of forming truth definitions, $Sat(A, B, \alpha)$ is defined

$$\exists C \Sigma(A, B, C, \alpha)$$

here Σ does not contain class variables[6] other than (the free variables) A, C.

Let U stand for the class of all sets, and E for the membership relation stricted to U.

[6] The definition has the following *invariance* property. If the set variables in Σ nge over a C_σ of the cumulative hierarchy, the classes are objects of $C_{\sigma+1}$, and the rticular formula above, for given A and B, defines the same set of α whether C nges over all of $C_{\sigma+1}$, or only over elements of $C_{\sigma+1}$ explicitly definable from A d B. The corresponding case for *higher*-order formulae is quite different.

For each particular $\tilde{\alpha}^1$ *we have:* $V\tilde{\alpha}^1 \to Sat(U,E,\tilde{\alpha})$ *provable in the theor of classes with axiom of infinity, hence* $V\tilde{\alpha}^1 \to \tilde{\alpha}_\epsilon$.
Cor. By a well-known result of Novak (see App. A[7]), $V\tilde{\alpha}^1 \to \tilde{\alpha}_\epsilon$ is provab in set theory for each formula $\tilde{\alpha}$.

The proof of the theorem uses $\forall\alpha(V\alpha^1 \to D_G\alpha^1)$ for *cut free* rules (e. Gentzen's), and then, for fixed $\tilde{\alpha}$, $D_G\tilde{\alpha} \to Sat(U,E,\tilde{\alpha})$ by means of a trut definition for *subformulae* of $\tilde{\alpha}$. For the proof of $\forall\alpha(V\alpha^1 \to D_G\alpha^1)$ on needs of course the axiom of infinity since some α are valid in all *fini* structures without being logically valid.

The machinery needed for this proof certainly justifies the reservatio above against the assumption that we simply *mean* V (i.e., truth in all s theoretic structures in some precise sense of set) when speaking of logic validity. Note incidentally, if we take any suitable *finitely* axiomatized s theory S, there is an $\tilde{\alpha}$ for which $V\tilde{\alpha} \to \tilde{\alpha}_\epsilon$ is not provable in S (namely, ta for $\tilde{\alpha}$ the negation of the conjunction of the axioms of S, granted that 'suitable' set theory cannot prove its own consistency, i.e., not $\sim V\tilde{\alpha}$).

The doubts are further confirmed by

(c) $\forall\alpha^1[V\alpha^1 \to Sat(U,E,\alpha^1)]$ *is not provable in the theory of classes.*

If it were, by the main result of App. A,[7] we should have an explicit s theoretic $\mathsf{F}\alpha^1$ such that

$$\forall\alpha^1[V\alpha \to \Sigma(U,E,\mathsf{F}\alpha,\alpha)]$$

is provable in the theory of classes. This reduces to a purely set-theore formula

$$(*) \qquad \forall\alpha^1[V\alpha \to \Sigma_1(\mathsf{F}\alpha,\alpha)]$$

which is proved in a finite subsystem S_1 of set theory; regarded as a form object of predicate logic (with $\in$ replaced by a binary relation symbol E let S_1 be σ_1 and let (*), regarded as a formal object, be π. Thus

$$\vdash_{S_1} V(\sigma_1 \to \neg\pi) \to \Sigma_1[\mathsf{F}(\sigma_1 \to \neg\pi), \sigma_1 \to \neg\pi].$$

But for the particular formula $\sigma_1 \to \neg\pi$, without use of induction, verify that $\langle U,E\rangle$ satisfies $\sigma_1 \to \neg\pi$, i.e.

$$\vdash_{S_1} \Sigma_1[\mathsf{F}(\sigma_1 \to \neg\pi), \sigma_1 \to \neg\pi] \to (S_1 \to \neg\forall\alpha[V\alpha \to \Sigma_1(\mathsf{F}\alpha,\alpha)]).$$

Since $\vdash_{S_1} S_1$, and, by assumption, $\vdash_{S_1}(*)$, we have $\vdash_{S_1} \neg V(\sigma_1 \to \neg\pi)$. B this would prove the consistency of S_1 in S_1.

Since *evidently on the intended interpretation of the theory of clas* (explicitly definable properties) (*) *is valid, we have found an instance*

7 [Omitted here.—Ed.]

-*incompleteness*. Thus looking at the intuitive relation *Val*, leads one not nly to formal proofs as in (a) but also to incompleteness theorems.

(d) All this was for first-order formulae. *For higher-order formulae we not have a convincing proof* of, e.g., $\forall\alpha^2(V\alpha \leftrightarrow Val\,\alpha)$ though one would :pect one. A more specific question can be formulated in terms of the erarchy of types C_σ. Let V^σ mean: truth in all structures that belong to C_σ. Then $\forall\alpha^1\forall\sigma > \omega(V^{\omega+1}\alpha \leftrightarrow V^\sigma\alpha)$ (Skolem–Löwenheim theorem). What the analogue (to ω) for second-order formulae? e.g., if $\tilde{\alpha}^2$ is Zermelo's stem of axioms, $V^{\omega+\omega+1}(\neg\,\tilde{\alpha}^2)$ is false, $V^{\omega+\omega}(\neg\,\tilde{\alpha}^2)$ is true. This analogue ω is certainly large. Let α assert of the structure $\langle a,e\rangle$ that (i) it is a C_σ r limit numbers σ, i.e., that $\langle a,e\rangle$ satisfies Zermelo's (second)-order ioms, (ii) $\langle a,e\rangle$ contains a measurable cardinal $>\omega$. Here (i) is of second der, and (ii) is of first order relative to (i). If $\beta = (\exists ae)\,\alpha$, we have $V^\sigma(\neg\,\beta)$ r $\sigma \leqslant$ the first measurable cardinal κ, but not $V^\sigma(\neg\,\beta)$ for $\sigma > \kappa$.

Since we do not even know a reduction analogous to the basic Skolem–oewenheim theorem, it is perhaps premature to ask for an analogue to $\alpha^1(V^{\omega+1}\alpha \leftrightarrow D\alpha)$. For instance, a well rounded theory of higher-order rmulae may be possible only for infinitely long ones. For infinite first-der formulae we do know an analogue when $D\alpha$ is replaced by certain *neralized inductive* definitions (cf. ω-rule).

General Conclusion. There is of course nothing new in treating *Val* as an nderstood concept; after all Gödel established completeness without aving to mention V; he simply used implicitly the obvious $\forall\alpha(Val\,\alpha \rightarrow {}^{\omega+1}\alpha)$ and $\forall\alpha(D\alpha \rightarrow Val\,\alpha)$ (incidentally for all i!), and proved $\alpha^1(V^{\omega+1}\alpha \rightarrow D\alpha)$. It seems a good time to examine this *solved* problem refully because (besides Heyting's rules for intuitionist validity, cf. ection 3) we face problems about finitist validity (Val_{F_i}) and predicative lidity (Val_P) not unlike those raised by Frege's rules. Thus, as in his case, e have (recursive) rules D_{F_i} and D_P for finitistic and predicative deducons respectively, established by means of autonomous progressions; and en equivalence to Val_{F_i}, resp. Val_P (for the languages considered) is most as plausible as was $Val \leftrightarrow D$ at the time of Frege. But we have not t found principles as convincing as those of Section 2(a) above to clinch e matter; in fact we do not have an analogue to V.

POSTSCRIPT

The following bibliographical information seems useful; incidentally, it plains the omission of those sections of the original article which are not printed here.

p. 79. Section 3 (omitted here) analyses Brouwer's empirical proposition which he uses in his proof of

$$\sim\forall\alpha[\sim\sim\exists x(\alpha x = 0) \supset \exists x(\alpha x = 0)].$$

These propositions have been further analysed by Kripke and Myhill [M]. The particular use of these propositions by Brouwer and analysed in Section 3 is superseded by a more delicate analysis of free choice sequence in paragraph 4 of [K2]. Incidentally, a much more interesting use of these propositions was made by Kripke who refuted, from them, Church's thesis by deriving $\sim\forall\alpha\, \exists p\, \forall n\, \exists m[T(p,n,m) \wedge an = Um]$. (Evidently we have $\sim\forall\alpha\, \exists p\, \forall n\, \exists m[T(p,n,m) \wedge \alpha n = Um]$ if α ranges over lawless sequences e.g., those generated by a die, since it would be absurd to *prove* of any die that it must follow a predetermined sequence. Kripke's argument is needed because it concerns a different notion of sequence.)

Appendix B is superseded by much stronger conservative extension results not only for non-standard arithmetic, but also for non-standard analysis [K3].

p. 86. Paragraph 4 of [K1] treats the step from second-order axioms to first-order schemata systematically.

p. 87. Second paragraph. A good theory of infinite formulae is developed in the dissertation of J. Barwise, Stanford University, 1967.

p. 93. Some additional information about $V^{\sigma}\alpha^2$ is contained on pp. 191– of [KK].

[K1] G. Kreisel, 'A Survey of Proof Theory', *Journal of Symbolic Logic*, Vol. 3 (1968), pp. 321–388.

[K2] G. Kreisel, 'Lawless Sequences of Natural Numbers', *Compositio*, Vol. 2 (1968).

[K3] G. Kreisel, 'Axiomatizations of Non-standard Analysis which are Conservative Extensions of Formal Systems for Standard Classical Analysis', to appear in *Applications of model theory to algebra, analysis and probability* (Holt, Rinehart an Winston).

[KK] G. Kreisel and J.-L. Krivine, *Elements of mathematical logic*, North-Hollan Publishing Co., Amsterdam, 1967.

[M] J. R. Myhill, 'Notes Towards an Axiomatization of Intuitionistic Analysis *Logique et Analyse*, Vol. 35 (1967), pp. 280–97.

VI

SYSTEMS OF PREDICATIVE ANALYSIS[1]

SOLOMON FEFERMAN

Dedicated to Professor Charles Loewner on the occasion of his 70th birthday.

THIS paper is divided into two parts. Part I provides a resumé of the evolution of the notion of predicativity.[2] Part II describes our own work on the subject.[3,4]

PART I

1. *Conceptions of Sets.* Statements about sets lie at the heart of most modern attempts to systematize all (or, at least, all known) mathematics. Technical and philosophical discussions concerning such systematizations and the underlying conceptions have thus occupied a considerable portion of the literature on the foundations of mathematics.

From these discussions we can distinguish fairly well at least the two extremes of what sets are conceived to be. In this section, we compare these two conceptions as they pertain to sets of finite type, in particular, sets of natural numbers. The remainder of the paper is devoted to systematizations of mathematical analysis based on one of these conceptions.

From one point of view, often identified as the *Platonistic* or *Cantorian* conception, sets have an existence which is independent of human definitions and constructions. The words 'arbitrary set' are often used to emphasize this independence. Various statements about sets are readily recognized to be correct under this conception, for example the axioms of

From the *Journal of Symbolic Logic*, Vol. 29 (1964), pp. 1–30. [Cf. also Introduction, p. 6 above.—Ed.]

[1] Text of an invited address delivered to a meeting of the Association for Symbolic Logic at Berkeley, California, on 26 January, 1963.

[2] For a recent interesting survey of this topic the reader should compare G. Kreisel, 'La prédicativité', *Bulletin de la Société Mathématique de France*, Vol. 88 (1960), pp. 371–91.

[3] We are much indebted to Professor Kreisel for many illuminating discussions about predicativity in general and our own work in particular.

[4] Much of the research for the new work described in Part II below was supported by grants from the U.S. Army Research Office [DA-ARO-(D)-31-124-G90 and 350].

comprehension and of choice. Other statements, such as the continuun hypothesis and its generalizations remain undecided on this conception (However, the inability of humans to decide such questions can no more b charged as a defect of this conception than can their inability to decid certain number-theoretical statements on the basis of the usual conception of the natural numbers.)

The other extreme is what we shall refer to as the *predicative* conception According to this, only the natural numbers can be regarded as 'given' to u (and, in the even more severe nominalist point of view, not even thes abstract objects are available to us). In contrast, sets are created by man t act as convenient abstractions (*façons de parler*) from particular condition or definitions. In order, for example, to predicatively introduce a set S o natural numbers x we must have before us a condition $\mathfrak{F}(x)$, in terms c which we define S by

$$(1.1)\qquad \wedge x[x \in S \leftrightarrow \mathfrak{F}(x)].$$

However, before we can assert the existence of such S, *it should already ha been realized that the defining condition $\mathfrak{F}(x)$ has a well-determined meanin which is independent of whether or not there exists a set S satisfying* (1.1) (bu which can depend on what sets have been previously realized to exist). I particular, to determing what members S has, we should not be led via $\mathfrak{F}(x$ into a *vicious-circle* which would return us to the very question we starte with. Conditions $\mathfrak{F}(x)$ which do so are said to be *impredicative*; it should b expected that most conditions $\mathfrak{F}(x)$ involving quantifiers ranging over 'arbi trary' sets are of this nature. Finally, we can never speak sensibly (in th predicative conception) of the 'totality' of all sets as a 'completed totality but only as a *potential totality* whose full content is never fully grasped bu only *realized in stages*.

The least upper bound principle of classical analysis provides an im portant example of the use of impredicative definitions. We can identif rational numbers with certain natural numbers and then identify real num bers with certain sets of natural numbers via Dedekind sections or Cauch sequences. If we identify real numbers with the upper parts of Dedekin sections we see that the l.u.b. of a (bounded, non-empty) set $\mathscr{M}$ of re numbers (with irrational l.u.b.) is simply given by $S = \bigcap X[X \in \mathscr{M}]$. If $\mathscr{M}$ i given by a condition $\mathfrak{G}(X)$,

$$(1.2)\qquad \wedge X[X \in \mathscr{M} \leftrightarrow \mathfrak{G}(X)],$$

then the intersection S is given by

$$(1.3)\qquad \wedge x[x \in S \leftrightarrow \wedge X(\mathfrak{G}(X) \rightarrow x \in X)].$$

ne existence of such S is, of course, clear under the Cantorian conception. owever, to answer the question 'What are the members of S?' we would, general, first have to know what sets X satisfy $\mathfrak{G}(X)$, and in particular hether or not $\mathfrak{G}(S)$ holds; this would, in turn, in general depend on iowing what members S has.[5]

Objections to the use of impredicative definitions in mathematics have en raised by a number of writers. At first, these objections centred on eir role in the paradoxes (Poincaré, Russell). However, it became clear fore long that the paradoxes were avoided by some simple restrictions on e use of the comprehension principle, while hardly abandoning impredi-tive definitions otherwise. The more thorough-going critics, such as Weyl, jected their use throughout mathematics and in particular as applied in e l.u.b. principle of analysis. Basically, these critics refused to believe that ere was any evidence to support the Cantorian conception of sets as dependently existing entities. Weyl described[6] impredicative analysis as house built on sand' in a 'logician's paradise'.

In defence of the Cantorian conception of sets we quote Gödel,[7] p. 137: seems to me that the assumption of such objects is quite as legitimate as e assumption of physical bodies and there is quite as much reason to lieve in their existence. They are in the same sense necessary to obtain a tisfactory system of mathematics as physical bodies are necessary for a tisfactory theory of our sense perceptions and in both cases it is imposs-le to interpret the propositions one wants to assert about these entities as opositions about the "data", i.e., in the latter case the actually occurring nse perceptions.' Earlier in the same essay (p. 127), Gödel supports, as rgely justified, the view that the axioms of logic and mathematics '... need t necessarily be evident in themselves, but rather their justification lies xactly as in physics) in the fact that they make it possible for these "sense rceptions" to be deduced; which of course would not exclude that they so have a kind of intrinsic plausibility similar to that in physics'. Presum-ly, the direct evidence of mathematics which is spoken of here and which rresponds to the 'sense perceptions' of physics includes at least finitisti-lly verifiable propositions concerning the natural numbers (which is not say that these are all the data).

The actual development of mathematics strongly supports one interpre-

[5] A specific example of such a condition which leads us into a vicious circle can be racted from the note at the end of this section.

[6] H. Weyl, *Das Kontinuum, Kritische Untersuchungen über die Grundlagen der alysis*, Leipzig (1918), iv + 84 pp.

[7] K. Gödel, 'Russell's mathematical logic', *The Philosophy of Bertrand Russell*, New rk, 1944, pp. 125–53.

tation of this argument. Abstraction and generalization are constantly pur-sued as the means to reach really satisfactory explanations which accoun for scattered individual results. In particular, extensive developments i algebra and analysis seem necessary to give us real insight into the behavic of the natural numbers. Thus we are able to realize certain results, whos instances can be finitistically checked, only by a detour via objects (such a ideals, analytic functions) which are much more 'abstract' than those wit which we are finally concerned.

The argument is less forceful when it is read as justifying some particula conceptions and assumptions, namely those of impredicative set theory, a *formally necessary* to infer the arithmetical data of mathematics. It is we known that a number of algebraic and analytic arguments can be system atically recast into a form which can be subsumed under elementary (fir order) number theory. According to the thesis and results of Kreisel,[8] suc arguments can be transformed even further into finitistically acceptab derivations, when the conclusions are of the form $\bigwedge x \bigvee y R(x,y)$ with primitive recursive. It is thus not at first sight inconceivable that predicativ mathematics is already (formally) sufficient to obtain the full range c arithmetical consequences realized by impredicative mathematics.

The only way to settle this is by a detailed investigation of the nature an potentialities of predicative reasoning. Such investigations began with th work of Russell on ramified analysis (which is described in the next section Following work of Weyl, Lorenzen, Schütte, Wang, Kleene, and Specto these investigations have resulted in two proposals, due to Kreisel and u as to what constitutes a *predicatively provable statement of analysis* (*give the natural numbers*). It is shown in Part II that these proposals are equi alent; we further obtain some fairly detailed information about the limi and potentialities of predicative reasoning in analysis. In particular, it shown that the arithmetical statement expressing the consistency of pred cative analysis is provable by impredicative means. Thus, if the these advanced below are granted, the situation envisioned at the end of th preceding paragraph is in fact not possible. Nevertheless, as will be ex plained in more detail below, it can be argued that *all mathematical interesting statements about the natural numbers, as well as many analyt statements, which have so far been obtained by impredicative methods ca already be obtained by predicative ones.*

Two further remarks are in order before we proceed to details. First of al it should be expected that the predicative conception of sets is comprehe

[8] G. Kreisel, 'Ordinal logics and the characterization of informal concepts proof', *Proceedings of the International Congress of Mathematicians* (1958), pp. 28 99.

sible to the Cantorian and that he should be able to consider whether a proposed explication of what constitutes predicative reasoning is both correct and complete. On the other hand, we cannot expect the confirmed predicativist to do any more than agree that any particular proposition coming under this heading is (meaningful and) correct. For, whenever he can recognize that all theorems of a certain set are correct, he can also recognize that the statement of consistency of this set is correct. However, the consistency of the set of all predicatively provable statements is not itself predicatively provable.

Second, it is not contended here that the two extremes we have considered are the only coherent or useful conceptions of sets. At least one intermediate conception of great interest is what is called the *constructive* conception. This permits, beside the use of predicative definitions, generalized inductive definitions of various sorts. It should be expected that this goes beyond the predicative, in the sense that we described following (1.1). For if we want to introduce the smallest set S satisfying certain *closure conditions* $\mathfrak{G}(S)$, we are in effect appealing to instances of the comprehension principle in the form given in (1.3) (the same form as used in the l.u.b. principle). (A specific example relevant to this argument is given in the note at the end of this section.) Several writers who have objected strongly to the full use of impredicative methods have also strongly urged restriction to constructive methods—in particular, Lorenzen[9] and Wang[10] have done so. Unfortunately, both these writers have used the term 'predicative' to refer to these more extensive notions as well (cf. also Lorenzen and Myhill,[11] and Wang,[12] Chs. XXIII–XXV, which contains the material of two papers of Wang's[13]). This is clearly a distortion of the original sense of predicativity as extracted from the Poincaré-Russell vicious-circle principle, which is the sense treated in this paper. Without denying the great interest and philosophical significance of constructivity, we feel that it is mistaken not to maintain the distinction drawn here.

NOTE. To see the impredicativity of a specific inductive definition, consider the usual such definition of the set O of constructive ordinal notations (cf.

[9] P. Lorenzen, 'Logical reflection and formalism', *Journal of Symbolic Logic*, Vol. 23 (1958), pp. 241–49.

[10] H. Wang, 'The formalization of mathematics', *Journal of Symbolic Logic*, Vol. 19 (1954), pp. 241–66.

[11] P. Lorenzen and J. Myhill, 'Constructive definition of certain analytic sets of numbers', *Journal of Symbolic Logic*, Vol. 24 (1959), pp. 37–49.

[12] H. Wang, *A Survey of Mathematical Logic*, Amsterdam, North-Holland Publishing Co., 1963.

[13] H. Wang, 'Ordinal numbers and predicative set theory', *Zeitschrift für mathematische Logik und Grundlagen der Mathematik*, Vol. 5 (1959), pp. 216–39. H. Wang, 'The formalization of mathematics', 1954, loc. cit.

Kleene[14]). This has the form $\bigwedge x[x \in O \leftrightarrow \bigwedge X(\mathfrak{A}(X) \rightarrow x \in X)]$, where $\mathfrak{A}$ is arithmetical and $\mathfrak{A}(O)$ holds. Suppose we were at a stage where we had predicatively realized the existence of all members of a certain collection $\mathscr{M}$, and that all members of $\mathscr{M}$ are hyperarithmetic. Then we could not use the formula $\mathfrak{F}(x) = \bigwedge X(\mathfrak{A}(X) \rightarrow x \in X)$ to predicatively determine the existence of the new set O at this stage, for the meaning of $\mathfrak{F}(x)$ will depend on whether or not we are including O in the range of the quantifier $\bigwedge X$. To see this, let $\mathscr{H}\mathscr{A}$ = the collection of hyperarithmetic sets and (impredicatively), $S_1 = \{x : \bigwedge X_{\mathscr{M}}(\mathfrak{A}(X) \rightarrow x \in X)\}$, $S_2 = \{x : \bigwedge X_{\mathscr{H}\mathscr{A}}(\mathfrak{A}(X) \rightarrow x \in X)\}$ and $O = \{x : \mathfrak{F}(x)\} = \{x : \bigwedge X_{\mathscr{M} \cup \{O\}}(\mathfrak{A}(X) \rightarrow x \in X)\}$. Then $S_1 \supseteq S_2 \supseteq O$; but $S_2 \in \sum_1^1$ by Kleene,[15] so $S_2 \neq O$ and hence also $S_1 \neq O$. For further discussion of these ideas, cf. Section 4 below.

2. *Ramified Analysis.* The first steps toward a formalization of predicative reasoning were made by Russell[16] by the introduction of type distinctions and within types of *order* distinctions or *ramifications* or, as we call them here, *degrees*.

The natural numbers are of type 0; we use letters $s, t, \ldots, x, y, z$ to range over these. Sets of natural numbers are of type 1; we use letters $S, T, \ldots, X, Y, Z$ for these. At type 2 we have collections $\mathscr{M}, \mathscr{N}, \ldots$ of sets of natural numbers, and so on. The $\in$ relation is to hold only between members of one type and those of the next type.

Let us restrict attention for the moment to types 0, 1, and to conditions $\mathfrak{F}(x)$ which involve variables just of these types. We call $\mathfrak{F}(x)$ *arithmetical* if it involves no quantification over type 1. Such conditions have perfectly well-determined meaning, given the natural numbers. We can introduce the totality $\mathscr{M}_0$ of sets S given by

$$\bigwedge x[x \in S \leftrightarrow \mathfrak{A}(x)] \tag{2.1}$$

for any arithmetical condition $\mathfrak{A}(x)$; we call these the sets of *degree* 0. Now if we consider an arbitrary condition $\mathfrak{F}(x)$, we can form a well-determined set S given by

$$\bigwedge x[x \in S \leftrightarrow \mathfrak{F}_{\mathscr{M}_0}(x)], \tag{2.2}$$

14 S. C. Kleene, 'Hierarchies of number-theoretic predicates', *Bulletin of the American Mathematical Society*, Vol. 61 (1955), pp. 193–213. 'On the forms of predicates in the theory of constructive ordinals', II, *American Journal of Mathematics*, Vol. 77 (1955), pp. 405–28.

15 S. C. Kleene, 'Quantification of number-theoretic functions', *Compositio Mathematica*, Vol. 14 (1959), pp. 23–40.

16 A. N. Whitehead and B. Russell, *Prinicipia Mathematica*, Vol. I, Cambridge University Press, 2nd ed., 1925.

where the subscript '$\mathscr{M}_0$' indicates relativization of all type 1 quantifiers in $\mathfrak{F}$ to $\mathscr{M}_0$. These sets are said to be of degree 1 and the totality of them is denoted by $\mathscr{M}_1$. We can proceed in this way to describe the sets of degree 2, 3, etc.

More generally, given any well-determined collection $\mathscr{M}$ and condition $\mathfrak{F}$, we shall write $\mathfrak{F}_{\mathscr{M}}$ for the result of relativizing all type 1 quantifiers in $\mathfrak{F}$ to $\mathscr{M}$, i.e. every occurrence $\bigwedge X(\ldots)$ in $\mathfrak{F}$ is replaced by $\bigwedge X_{\mathscr{M}}(\ldots)$, i.e., $\bigwedge X(X \in \mathscr{M} \rightarrow \ldots)$, (and similarly for existential quantifiers). Then we introduce:

(2.3) *$\mathscr{M}^*$ consists of all sets S such that for some condition $\mathfrak{F}(x)$, we have* $\bigwedge x[x \in S \leftrightarrow \mathfrak{F}_{\mathscr{M}}(x)]$.

This procedure described in the preceding paragraph consists of introducing the collections $\mathscr{M}_k$ by:

(2.4) (i) *$\mathscr{M}_0$ consists of all arithmetically definable sets*;
(ii) $\mathscr{M}_{k+1} = \mathscr{M}_k^*$.

This semantical description (in terms of definability) is the intended model of a formal system for types 0, 1. Instead of one kind of variable for type 1, we now have different variables $S^k, T^k, \ldots, X^k, Y^k, Z^k$ for each degree k. By the *degree of a formula* $\mathfrak{F}$ in this system we mean 0, if it contains no free or bound variables of type 1, otherwise the maximum of all $m+1$ such that a variable X^m appears bound in $\mathfrak{F}$, and of all n such that a variable Y^n appears free in $\mathfrak{F}$. Then the *comprehension axiom schema* takes the form

(2.5) $$\bigwedge S^k \bigwedge x[x \in S^k \leftrightarrow \mathfrak{F}(x)]$$

for each formula $\mathfrak{F}$ of degree k (in which S^k does not appear free). This is the essential feature of Russell's *ramified theory of types*. It certainly accords with part of the predicative conception of sets.

In the above theory we have no way of talking about 'arbitrary' real numbers (as, say, Dedekind sections). We can only talk about real numbers of degree 0, of degree 1, of degree 2, etc. Russell realized this would be ludicrous to practising analysts. Although he expressed great reservations concerning the following step, he found it necessary to introduce the *axiom(s) of reducibility*:

(2.6) $$\bigwedge S^n \bigvee S^0 \bigwedge x[x \in S^0 \leftrightarrow x \in S^n],$$

in effect wiping away the system of degrees. It took Ramsey to finally observe that the system of types already avoided the paradoxes, to which the detour through degrees and back again via the axiom of reducibility added nothing. Thus analysis as analysts know it could be developed in the *simple theory of types* without the obvious paradoxes being forthcoming.

3. *Toward the Practice of Predicative Analysis.* In his work *Das Kontinuum*[17] of 1918, Weyl took the resolute and well-needed step toward seeing exactly how much remained in a system of analysis built on strictly predicative foundations. Weyl also saw that analysis with real numbers of different degrees was unacceptable; he thus decided to disregard all sets above degree 0, i.e., above the arithmetically definable sets. In modern formal terms (and now showing explicitly the possible use of additional free variables in defining conditions) his comprehension axioms were just those of the form:

$$(3.1)\quad \bigwedge y, z, \ldots \bigwedge Y, Z, \ldots \bigvee S \bigwedge x[x \in S \leftrightarrow \mathfrak{A}(x, y, z, \ldots, Y, Z, \ldots)],$$

where $\mathfrak{A}$ is any arithmetical formula in which the variable 'S' does not appear. We shall call the (second order) system of analysis built with this restricted principle the system of *arithmetic* or *elementary analysis*. It should be observed that this system can be regarded as a subsystem of both ramified and classical analysis.

The surprising result found by Weyl was that essentially the whole analysis of continuous functions as contained, say, in the standard undergraduate courses could be developed in this system. We have, for example, for a continuous function on a closed interval of real numbers, existence of maximum and minimum, mean value theorem, uniform continuity, existence of Riemann integral, and the Fundamental Theorem of Calculus.

These results about continuous functions depend among other things on the Bolzano-Weierstrass and Heine-Borel Theorems. How is it possible to obtain these without the l.u.b. principle? The answer is that in the cases of interest we can deal with l.u.b. of *sequences*, rather than sets, and in fact sequences of a certain special form. If we identify real numbers with (upper) Dedekind sections in the rationals, we say that a sequence S_n of real numbers is *given* if we have a set T such that for each n, $x \in S_n \leftrightarrow \langle x, n\rangle \in T$. Then the l.u.b. of the S_n, which we identified earlier with $\bigcap S_n[n \in \omega]$, is defined arithmetically in terms of T by $\bigwedge y(\langle x, y\rangle \in T)$. It is in this sense that *we can prove in elementary analysis that every bounded sequence of reals has a least upper bound.* (There is no essential difference here if real numbers are represented by Cauchy sequences in the rationals.) The corresponding version

[17] See p. 97, note 6.

of the Bolzano-Weierstrass theorem, in the form that every bounded sequence of reals contains a convergent subsequence, is then provable by the usual subdivision process. Similarly, we can prove the Heine-Borel theorem when expressed in the form: *Given a sequence* (s_n, t_n) *of open rational intervals, if for each n,* $\bigcup_{i \leq n}(s_n, t_n)$ *does not cover* $[a,b]$ *then there exists a real number* $r \in [a,b] - \bigcup_{n=0}^{\infty}(s_n, t_n)$. The number r is arithmetically definable in terms of the given sequence.

The carry-over of the above ideas to the complex numbers and the classical theory of analytic functions is then straightforward. It is also clear that all the usual special functions of classical analysis, including the various algebraic and transcendental functions, are arithmetically definable, as seen from their power series expansions.

The ideas of Weyl languished for some time, but they have received renewed interest in recent years. These ideas were carried out in modern form and in some detail by Grzegorczyck[18] and, in another form, by M. Kondô.[19] Beyond this, Kreisel[20] has observed that somewhat further portions of real analysis can be developed arithmetically (but allowing higher type variables), for example, parts of the theory of Lebesgue measurable sets and functions. It is not claimed that the full notion of (outer) measure can be introduced, for this essentially involves the notion of g.l.b. However, it is a consequence of the usual definitions of measurability that a set $\mathscr{M}$ of real numbers in $(0,1)$ is measurable if and only if there is a G_δ set $\mathscr{N} \supseteq \mathscr{M}$ such that $\mu^*(\mathscr{N} - \mathscr{M}) = 0$ (outer measure); equivalently that there are G_δ sets $\mathscr{N}_1 \supseteq \mathscr{M}$, $\mathscr{N}_2 \supseteq [0,1] - \mathscr{M}$ with $\mu^*(\mathscr{N}_1) + \mu^*(\mathscr{N}_2) = 1$. Now we would say that a G_δ set $\mathscr{N}$ is given if we have a double sequence of pairs $\langle s_{nm}, t_{nm} \rangle$ of rationals such that $\mathscr{N} = \bigcap_{n=0}^{\infty} \bigcup_{m=0}^{\infty}(s_{nm}, t_{nm})$. Then $\mu^*(\mathscr{N})$ is arithmetically definable in terms of this sequence. We explain similarly what is meant by giving a sequence of G_δ sets. Then we can deal in elementary analysis with measurable sets or sequences of such for which corresponding G_δ sets (as above) or sequences of such are given. In this way a good deal of classical analysis can already be recaptured in elementary analysis.

This is of interest in two ways. With respect to questions concerning the foundations of mathematics, it begins to bear out the contention of Section 1 that only predicative concepts need by assumed in order to derive all known

18 A. Grzegorczyck, 'Elementary definable analysis', *Fundementa Mathematica*, Vol. 41 (1955), pp. 311–38.

19 M. Kondô, 'Sur les ensembles nommables et le fondement de l'analyse mathematique', I, *Japanese Journal of Mathematics*, Vol. 28 (1958), pp. 1–116.

20 G. Kreisel, 'The axiom of choice and the class of hyperarithmetic functions', *Koninklijke Nederlands Akademie van Wetenschappen, Proceedings*, ser. A, Vol. 65 (also *Indagationes Mathematicae*, Vol. 24) (1962), pp. 307–19.

arithmetical results and a number of analytical results which are of mathematical interest. On the other hand, it isolates various closure conditions on a collection of real numbers which are necessary to obtain such results, in this case closure under arithmetical definability. This points to a different kind of generalization of analysis than has been pursued by analysts up to now, and one which is perhaps more related to the kind followed by algebraists.

What is still lacking at this point is an adequate explication of the full content of predicative principles. Such an explication should somehow get around the problem of dealing with real numbers of different degrees; it should also give us greater freedom to deal with objects of higher types.

4. *Toward the Formalization of Predicative Analysis.* The first step to obtain the desired explication of the notion, *predicatively provable statement of (classical) analysis*, is to obtain a better understanding of the notion, *predicatively definable set of natural numbers.* This suggests a reconsideration of the collections $\mathscr{M}_k$ defined in (2.4). If we accept that each member of each $\mathscr{M}_k$ is predicatively definable, then we should also accept this for $\bigcup \mathscr{M}_k[k < \omega]$. This collection, which we denote by $\mathscr{M}_\omega$, is predicatively well-determined. Hence every member of $\mathscr{M}_{\omega+1} = \mathscr{M}^*_\omega$ is predicatively definable, and then so is every member of $\mathscr{M}_{\omega+2} = \mathscr{M}^*_{\omega+1}$, etc. In general, we define a transfinite sequence of collections $\mathscr{M}_\alpha$ by:

(4.1) (i) $\mathscr{M}_0$ *consists of all arithmetically definable sets,*
(ii) $\mathscr{M}_{\alpha+1} = \mathscr{M}^*_\alpha$ *for each* α, *and*
(iii) $\mathscr{M}_\alpha = \bigcup \mathscr{M}_\beta[\beta < \alpha]$ *for limit* α.

Now we must make one important restriction on this procedure. In the predicative conception, the notion of 'arbitrary' ordinal, i.e., 'arbitrary' well-ordering relation, is just as meaningless as the notion of 'arbitrary' set. Each ordinal used is to be determined by a well-ordering relation $x < y$ and this relation, considered as a set of ordered pairs $\langle x, y\rangle$ must already be among the admitted sets. This leads to the following *proviso of autonomy* on (4.1):

(4.2) *In* (4.1) (iii) *we consider only those limit ordinals* α *for which there is some* $\gamma < \alpha$ *and some set* S *of ordered pairs* $\langle x, y\rangle$ *such that*:
(i) $S \in \mathscr{M}_\gamma$ *and*
(ii) S *is a well-ordering of its field of order type* α.

Without the restriction (4.2), the sequence defined in (4.1) gives the second-order analogue of the *constructible sets* of natural numbers defined by

Gödel.[21] While this notion is of great mathematical interest, its distinction from a more strictly predicative procedure should be kept in mind.[22]

The proviso (4.2) suggests introducing, instead, collections $\mathscr{M}_<$ associated with well-ordering relations $x < y$. It follows from the work of Spector[23] (via results of Kleene[24]) that the resulting collections $\mathscr{M}_<$ would depend only on the order type $\alpha = |<|$ of the given ordering. Furthermore, Spector showed that:

(4.3) *we can already restrict ourselves in* (4.2) *to well-orderings* $S \in \mathscr{M}_0$, *and in fact to recursive* S.

The least ordinal not thus obtained is denoted by ω_1.

Now the following thesis has been proposed by Kreisel (on several occasions, one in particular[25]):

(4.4) *every predicatively definable set belongs to* $\mathscr{M}_{\omega_1}$.

This thesis seems to us to be well-grounded on our basic conceptions of the notion in question. Whether the converse thesis, that *every member of* $\mathscr{M}_{\omega_1}$ *is predicatively definable*, should also be accepted is more debatable. In any case, the definition of $\mathscr{M}_{\omega_1}$ as given is clearly impredicative; we might say that $\mathscr{M}_{\omega_1}$ is what the predicative 'universe' looks like to the impredicativist. Since our main concern here is with the notion of predicative provability rather than definability, a defence of the thesis (4.4) and/or its converse need not be pursued. However, information about $\mathscr{M}_{\omega_1}$ will turn out to be extremely useful as a heuristic guide.

A very important result about $\mathscr{M}_{\omega_1}$ has been obtained by Kleene:[26]

(4.5) $\mathscr{M}_{\omega_1}$ *consists exactly of the hyperarithmetic sets.*

21 K. Gödel, *The consistency of the axiom of choice and the generalized continuum hypothesis with the axioms of set theory*, Annals of Mathematics Studies No. 3, Princeton (1940).

22 Writing of his work on constructible sets (in 'Russell's mathematical logic', 1944, loc. cit.), Gödel says: 'The theory of orders [i.e., ramification degrees] proves more fruitful if considered from a purely mathematical standpoint, independently of the philosophical question whether impredicative definitions are admissible. Viewed in this manner, i.e., as a theory built up within the framework of ordinary mathematics, where impredicative definitions are admitted, there is no objection to extending it to arbitrarily high transfinite orders. ...[then] all impredicatives are reduced to one special kind, namely the existence of certain large ordinals (or, well ordered sets) and the validity of recursive reasoning for them.'

23 C. Spector, 'Recursive well-orderings', *Journal of Symbolic Logic*, Vol. 20 (1955), pp. 151–63.

24 S. C. Kleene, 'Quantification of number-theoretic functions', 1959, loc. cit.

25 G. Kreisel, 'La predicativité', 1960, loc. cit.

26 S. C. Kleene, 'Quantification of number-theoretic functions', 1959, loc. cit.

These sets, the study of which was initiated by Mostowski, and was then carried on more intensively by Kleene and Spector (among others), have been shown by Kleene[27] to coincide with the sets in both one-analytic-quantifier forms, i.e.:

(4.6) *a set S is hyperarithmetic if and only if there are arithmetical conditions* $\mathfrak{A}(x, Y)$, $\mathfrak{B}(x, Z)$ *such that*

(i) $\bigwedge x[\bigvee Y\mathfrak{A}(x, Y) \leftrightarrow \bigwedge Z\mathfrak{B}(x, Z)]$

and

(ii) $\bigwedge x[x \in S \leftrightarrow \bigvee Y\mathfrak{A}(x, Y)]$.

In general, unrestricted set-theoretical quantification of the sort given in (4.6) is predicatively meaningless. However, the following more detailed information obtained by Kleene[28] permits us to ascribe well-determined sense to such quantifications.

(4.7) *We can find collections* $\mathscr{H}_\beta$, $\beta < \omega_1$, *with the following properties*:

(i) $\mathscr{M}_\alpha = \mathscr{H}_{\omega(1+\alpha)}$ *for each* $\alpha < \omega_1$,
(ii) *for each limit* $\beta < \omega_1$, $\mathscr{H}_\beta = \bigcup \mathscr{H}_\gamma[\gamma < \beta]$, *and*
(iii) *for each* $\beta < \omega_1$ *and* $S \in \mathscr{H}_{\beta+1}$, *we can find arithmetic conditions* $\mathfrak{A}$, $\mathfrak{B}$ *satisfying both* (4.6) (i), (ii) *as given and also when all quantifiers are relativized to* $\mathscr{H}_\beta$—*hence also satisfying* (4.6) (i), (ii) *when the quantifiers are relativized to any* $\mathscr{N}$ *for which* $\mathscr{H}_\beta \subseteq \mathscr{N}$.[29]

In other words, the reference to the 'completed totality' of all sets in the quantifications of (4.6) is only an apparent one, and one can read these equally well and *unambiguously* as referring to a 'potential totality', of which a well-determined part $\mathscr{H}_\beta$ is already before us.

This leads to another argument in support of (4.4). Given a collection $\mathscr{M}$ and an arbitrary second-order formula $\mathfrak{F}(x)$ with just x free, we say that $\mathfrak{F}$ *is definitive relative to* $\mathscr{M}$ if for every $\mathscr{N} \supseteq \mathscr{M}$ we have $\bigwedge x\{\mathfrak{F}_{\mathscr{M}}(x) \leftrightarrow \mathfrak{F}_{\mathscr{N}}(x)\}$. In particular, if $\mathfrak{F}$ is definitive relative to $\mathscr{M}$ then $\bigwedge x\{\mathfrak{F}_{\mathscr{M}}(x) \leftrightarrow \mathfrak{F}(x)\}$. Then we introduce:

(4.8) $\mathscr{M}^+$ *consists of all members of* $\mathscr{M}$ *together with all sets* S *such that for some formula* $\mathfrak{F}(x)$ *which is definite relative to* $\mathscr{M}$, *we have* $\bigwedge x[x \in S \leftrightarrow \mathfrak{F}_{\mathscr{M}}(x)]$.

27 S. C. Kleene, 'Hierarchies of number-theoretic predicates', 1955, loc. cit.
28 S. C. Kleene, 'Quantification of number-theoretic functions'.
29 In terms of the hyperarithmetic hierarchy of Kleene's 'Hierarchies of number-theoretic predicates', $\mathscr{H}_\beta$ is defined to consist of all sets recursive in some H_c for $|c| < \beta$.

This leads us to define a transfinite sequence of collections $\mathscr{D}_\alpha$ by:

(4.9) (i) *$\mathscr{D}_0$ consists of all arithmetically definable sets,*
(ii) *$\mathscr{D}_{\alpha+1} = \mathscr{D}_\alpha^+$ for each α, and*
(iii) *$\mathscr{D}_\alpha = \bigcup \mathscr{D}_\beta[\beta < \alpha]$ for limit α.*

Since $\bigcup \mathscr{D}_\alpha$, α arbitrary, contains only analytically definable sets, it is denumerable. Hence if Ω is the first non-denumerable ordinal, the process must stop by Ω, i.e., $\mathscr{D}_\Omega = \mathscr{D}_{\Omega+1}$. In fact we have

(4.10) *$\mathscr{D}_\Omega = \mathscr{D}_{\omega_1}$ = the collection of all hyperarithmetic sets.*

The proof in one direction, that every hyperarithmetic set belongs to $\mathscr{D}_{\omega_1}$ makes use of the result of Kleene's stated in (4.7). We now show (4.10) in the other direction, namely that each element of $\mathscr{D}_\Omega$ is hyperarithmetic. Let $\mathscr{N}$ be any collection which provides an ω-model of classical analysis, i.e., for any formula $\mathfrak{F}$, $\{x : \mathfrak{F}_{\mathscr{N}}(x)\} \in \mathscr{N}$. We prove by induction on α that $\mathscr{D}_\alpha \subseteq \mathscr{N}$. It is sufficient to show that if this holds for $\mathscr{D}_\alpha$ it also holds for $\mathscr{D}_{\alpha+1}$. Given any new element S of $\mathscr{D}_{\alpha+1}$, we have $S = \{x : \mathfrak{F}_{\mathscr{D}_\alpha}(x)\}$ for $\mathfrak{F}(x)$ definite relative to $\mathscr{D}_\alpha$, hence also $S = \{x : \mathfrak{F}_{\mathscr{N}}(x)\}$ and thus $S \in \mathscr{N}$. We can thus conclude that for any such $\mathscr{N}$, $\mathscr{D}_\Omega \subseteq \mathscr{N}$. Put in slightly different terms, $\mathscr{D}_\Omega \subseteq \bigcap \mathscr{N}$ [$\mathscr{N}$ is an ω-model of classical analysis]. However, by Theorem · of Gandy, Kreisel and Tait,[30] this intersection contains only hyperarithmetic sets. (This argument and the result (4.10) can be improved further by restricting the range of the quantifier 'for all $\mathscr{N}$' in the explanation of the notion of definiteness.)

The sequences $\mathscr{M}_\alpha$ and $\mathscr{D}_\alpha$ provide two ways of looking at predicative definability, realized by two ways of turning a rejection of impredicative definitions into a criterion as to which definitions are acceptable. These then correspond, roughly, to two approaches to formalizing predicative analysis. The sequence $\mathscr{M}_\alpha$ corresponds to the ramified approach, in which the instances of the comprehension axiom to be used are restricted by attaching degree indices to all variables of type 2 and restricting the relationships between these indices; however, the formulas used as definitions are otherwise quite arbitrary. The sequence $\mathscr{D}_\alpha$ suggests the possibility of an unramified formalization in which the conditions to be used in the comprehension axiom satisfy a condition of definiteness.

A return to 'constructive' mathematics via a transfinite ramified theory, among other devices, has been particularly urged during the last few years

30 R. Gandy, G. Kreisel, and W. Tait, 'Set existence', *Bulletin de l'Académie Polonaise des Sciences*, Série des sciences mathématiques, astronomiques et physiques, vol. 8 (1960), pp. 577–82.

by Lorenzen[31] and Wang.[32] Despite the philosophical attractiveness of their position, they did not succeed in establishing it definitively. First of all, the exact nature of their proposals has never been made completely clear, even taking a broad view of the nature of formalization.[33] In particular, these authors did not show how the transfinitely conceived sequence $\mathscr{M}_\alpha(\alpha < \omega_1)$ is to be dealt with in a system of proofs, each of which consists of finitely many symbols. Secondly, and more importantly, they failed to meet the old objection that ramified analysis is a parody of classical analysis, not a significant part of it.

This objection was finally met by Kreisel, who recently made[34] a new proposal for explicating the basic rules of predicative reasoning. This was suggested to him by a return to the idea of 'potential totalities' which we have discussed in the foregoing in terms of definite formulas. The rule suggested by Kreisel is (for $\mathfrak{A}$, $\mathfrak{B}$ arithmetical):

(4.11) *if we have inferred* $\bigwedge x[\bigvee Y\mathfrak{A}(x,Y) \leftrightarrow \bigwedge Z\mathfrak{B}(x,Z)]$, *then we are allowed to infer* $\bigvee S \bigwedge x[x \in S \leftrightarrow \bigvee Y\mathfrak{A}(x,Y)]$.

Because of its relation to (4.6), we call this the *hyperarithmetic comprehension rule* (HCR). Compared to deductions in ramified analysis, every statement proved using (HCR) has immediate significance in classical analytic terms and, in fact, is a theorem of classical analysis. Furthermore, we have a predicative justification of (4.11), which is the proof-theoretic analogue of definiteness. This was given by Kreisel;[35] we repeat it here.

(4.12) *With each statement* $\mathfrak{F}$ *in a proof using* (HCR) *we can associate a collection* $\mathscr{M}$ *such that* $\mathfrak{F}_{\mathscr{N}}$ *is correct no matter what* $\mathscr{N}$ *we take with* $\mathscr{M} \subseteq \mathscr{N}$.

For suppose we have such $\mathscr{M}$ associated with the hypothesis of an (HCR) as in (4.11). Suppose that there are no free set variables in this hypothesis. Then we have

$$\bigwedge x[\bigvee Y_{\mathscr{M}}\mathfrak{A}(x,Y) \leftrightarrow \bigwedge Z_{\mathscr{M}}\mathfrak{B}(x,Z)]$$

and

$$\bigwedge x[\bigvee Y_{\mathscr{N}}\mathfrak{A}(x,Y) \leftrightarrow \bigwedge Z_{\mathscr{N}}\mathfrak{B}(x,Z)]$$

31 P. Lorenzen, 'Logical Reflection and Formalism', 1958, loc. cit.
32 H. Wang, *A Survey of Mathematical Logic*, op. cit.
33 Wang himself repeatedly describes his attempts at formalization in *A Survey of Mathematical Logic*, Chs. XXIII–XXV, as being of a tentative nature.
34 G. Kreisel, 'The axiom of choice and the class of hyperarithmetic functions', 1962, loc. cit.
35 Ibid.

or any $\mathcal{N} \supseteq \mathcal{M}$. Thus we conclude that

$$\wedge x[\vee Y_{\mathcal{N}} \mathfrak{A}(x, Y) \leftrightarrow \vee Y_{\mathcal{M}} \mathfrak{A}(x, Y)].$$

Hence if we take $\mathcal{M}_1$ to consist of $\mathcal{M}$ and $\{x : \vee Y_{\mathcal{M}} \mathfrak{A}(x, Y)\}$, we see that

$$\vee S_{\mathcal{N}} \wedge x[x \in S \leftrightarrow \vee Y_{\mathcal{N}} \mathfrak{A}(x, Y)]$$

now holds for every $\mathcal{N} \supseteq \mathcal{M}_1$.[36]

It is (4.12) which allows us to make use of formally unrestricted set quantifiers in a development of predicative analysis, where this would otherwise be illegitimate.

PART II

We are now prepared to describe the general nature of our own contributions in this subject, and which are to be explained more fully below.[37] Unfortunately, space does not permit us to give anything more than a hint of the proofs, some of which have turned out to be rather long. A detailed account of these is now being prepared and will appear elsewhere.

The starting point consists of two proposed formal explications of the notion *predicative provability in* (*second order*) *analysis*, both of which fall under the general frame of *autonomous progressions of theories*.[38] The first of these (originally suggested by Kreisel[39]) is in terms of a transfinite sequence of theories R_α embodying transfinite ramified analysis, while the second makes use of a sequence of theories H_α embodying the rule (HCR), as well as a formalized ω-rule. These proposals are described more precisely in Section 5 and Section 6.

It turns out that these are essentially equivalent formalizations. By this we mean that certain mutual intertranslatability results hold; these are described in more detail in Section 6. Since the given formalizations arose from two different ways of looking at predicativity, the equivalence supports

36 To account for the general case that free set variables appear in the hypothesis of an application of (HCR), which is not excluded in (4.11), the statement of (4.12) must be modified slightly so that we deal with $\mathcal{M}$ satisfying certain closure conditions and only those $\mathcal{N} \supseteq \mathcal{M}$ which satisfy the same closure conditions. The preceding argument is then modified accordingly.

37 Part of this work has been reported in two abstracts in the *Notices of the American Mathematical Society*: 'Constructively provable well-orderings' (Vol. 8 (1961), p. 495) and 'Provable well-orderings of and relations between predicative and ramified analysis', Vol. 9 (1962), p. 323.

38 S. Feferman, 'Transfinite recursive progressions of axiomatic Theories', *Journal of Symbolic Logic*, Vol. 27 (1962), pp. 259–316; G. Kreisel, 'Ordinal logics and the characterization of informal concepts of proof', 1958, loc. cit.

39 G. Kreisel, 'The axiom of choice and the class of hyperarithmetic functions', 1962, loc. cit.

the thesis that we are dealing with the correct explication of the notion of predicative provability. We show further that the sequence of theories H_α is closed under a certain *rule of choice*, which is *prima facie* stronger than (HCR), and which was also advanced by Kreisel[40] as being a correct predicative rule. Of special interest in these arguments is the identification of the *least non-predicatively provable well-ordering*, whose ordinal is denoted by Γ_0. Some of this work was also obtained independently by Schütte.[41]

We have gone on in Section 6 to isolate a single system, which we refer to as IR, and which has exactly the same theorems as $\bigcup H_\alpha$ $(\alpha < \Gamma_0)$. In addition to (HCR), this makes use of certain rules permitting transfinite induction and recursion.

Finally, we indicate in Section 7 how to obtain a smooth extension of the theories H_α to arbitrary higher types. We examine in the light of this extension the newly available instances of the l.u.b. principle and indicate to what additional extent we are capable of predicatively developing significant parts of classical analysis.

5. *Well-orderings, Ordinal Notations, and Progressions of Theories.* Well orderings and ordinal notations enter essentially in the following, not only as a means for dealing with sequences of theories, but also for the main arguments which bring out the content of these theories.

All ordinals to be considered here are denumerable. We are especially interested in the subset of these which are order types of recursive well orderings of natural numbers. A convenient system of representatives of these is given by the set O of constructive ordinal notations first developed by Church and Kleene. We assume some familiarity with this.[42] The system is taken in slightly modified form here. We have a recursively enumerable relation $a < b$ such that: for all a, b,

(i) $a \not< 0$,
(ii) $a < 2^b \leftrightarrow a < b \vee a = b$,
(iii) $a < 3 \cdot 5^b \leftrightarrow (En)(a < [b]_n)$.

Here $[b]$ is the bth primitive recursive function in a standard enumeration and $[b]_n = [b](n)$. We then have O as the smallest set which contains 0, and

40 G. Kreisel, 'The axiom of choice and the class of hyperarithmetic functions', 1962 loc. cit.

41 K. Schütte, 'Eine Grenze fur die Beweisbarkeit der transfiniten Induktion in der verzweigten Typenlogik', to appear.

42 Cf., e.g., Kleene, 'On the forms of predicates in the theory of constructive ordinals', 1955, loc. cit., or S. Feferman, 'Classifications of recursive functions by means of hierarchies', *Transactions of the American Mathematical Society*, Vol. 104 (1962), pp. 101–122.

uch that whenever it contains b it also contains 2^b, and whenever it contains $b]_n$ for each n and $[b]_n < [b]_{n+1}$ for all n, it also contains $3{\cdot}5^b$. The set O is artially ordered by $<$, with the set of predecessors of each $b \in O$ well-rdered by $<$. For $b \in O$, $|b|$ denotes the corresponding ordinal. Then ω_1 is he least ordinal not so denoted. We have $|0| = 0$, $|2^b| = b+1$ and $3{\cdot}5^b| = \sup_n |[b]_n|$.

By means of the recursion theorem one can define various primitive ecursive functions which correspond on O to functions of ordinals given y transfinite recursions. We have, for example, such functions $a \oplus b$, $\circ\, b$, $a^{\circ b}$, corresponding to ordinal addition, multiplication and expo-entiation, so that $|a \oplus b| = |a| + |b|$, etc.[43] In particular $a \oplus 1 = 2^a$. While nese are adequate for many purposes, in other respects they are limited. 'or example $\oplus$ is not associative; also while $a \leq a \oplus b$, in general we do ot have $b \leq a \oplus b$. These limitations are disadvantageous when one wants) deal with critical numbers of functions. For example, classically we lways have $\beta \leqq \omega^\beta$ and the ϵ-numbers are the solutions β of the equation $^\beta = \beta$. However, if we take w with $|w| = \omega$, in general we do not have $\leq w^{\circ b}$, and we can't speak directly of a function ϵ_a enumerating the olutions of $|w^{\circ b}| = |b|$ in O. Since we shall have to make systematic use of ffective versions of various kinds of critical numbers and related functions, e must also consider a narrower system than O.

We call a set A of ordinals *closed* if it is closed under limits. A function from ordinals to ordinals is *continuous* if $F(\alpha) = \sup_{\beta<\alpha} F(\beta)$ for every mit α. We call F *normal* if it is continuous and strictly increasing. The nge of every normal function F is a closed and unbounded set A; con-ersely, every such A is the range of a normal function. For F strictly creasing, we always have $\beta \leqq F(\beta)$. Then $\{\beta : F(\beta) = \beta\}$ is closed and un-ounded if, in addition, F is normal. We call this the set of *critical points* of . It is enumerated by a normal function F', which we also call *the critical nction* of F.

Suppose $F^{(\nu)}$, $\nu < \tau$, is a sequence of normal functions with ranges $A^{(\nu)}$, nd that $A^{(\mu)} \subseteq A^{(\nu)}$ for $\mu > \nu$. Then $\bigcap A^{(\nu)}[\nu < \tau]$ is also closed and nbounded. We can then associate with this sequence a function $F^{(\tau)}$ numerating this intersection; it is called *the critical function of the sequence* $^{(\nu)}(\nu < \tau)$.

Now if we start with a given normal function $F^{(0)}$ we can form a trans-nite sequence of normal functions as follows:

.1) (i) *$F^{(0)}$ is given*;

43 Cf. Feferman, 'Classifications of recursive functions by means of hierarchies' 962, loc. cit.).

(ii) $F^{(\nu+1)} = (F^{(\nu)})'$, *i.e.*, $F^{(\nu+1)}$ *is the critical function of* $F^{(\nu)}$;

(iii) *for* τ *a limit number*, $F^{(\tau)}$ *is the critical function of the sequenc* $F^{(\nu)}(\nu < \tau)$.

For example, if we start with $F^{(0)}(\alpha) = \omega^{\alpha}$, and in general write $\chi_{\alpha}^{(\nu)}$ fo $F^{(\nu)}(\alpha)$, we have:

(5.2) (i) $\chi_{\alpha}^{(0)} = \omega^{\alpha}$;

(ii) $\chi_{\alpha}^{(\nu+1)}$ = *the* α^{th} *solution* β *of* $\chi_{\beta}^{(\nu)} = \beta$;

(iii) *for* τ *a limit number* $\chi_{\alpha}^{(\tau)}$ = *the* α^{th} *solution* β *of* $\chi_{\beta}^{(\nu)} = \beta$ *for a* $\nu < \tau$.

If can then be seen that if we set

(5.3) $$\gamma_{\beta} = \chi_0^{(\beta)}$$

then

(5.4) γ_{β} *is a normal function of* β, *and we have a normal function* Γ_{α} *whic enumerates its critical points.*

Using only the starting 'symbol' 0, and writing $a \oplus b$ and $\chi_a^{(b)}$ as ne 'symbols' whenever a, b are 'symbols', we can denote a substantial segmen of ordinals. Namely, the given symbolism will assign a notation to ever ordinal $<\Gamma_0$ (and only such ordinals), and this notation can be chosen in unique way by means of an effective procedure.

These ideas are the start for developing 'natural' systems of notation fo larger and larger segments of the ordinals. A considerable extension of th above procedure was first presented by Veblen,[44] and has since been recas into modern form by Schütte.[45] By Schütte's work, we know that th ordering relation in this system of unique notations is primitive recursiv Further, we have a uniform primitive recursive procedure for assigning t each limit notation b a primitive recursive fundamental sequence b_n. follows by standard arguments that this system of notations can be effe tively embedded in O. We denote the resulting subset of O by O_N ('N' fo 'natural'). It can also be arranged that O_N is closed under the $\oplus$ of O an that we have primitive recursive functions $\chi_a^{(b)}$ on O_N with $|\chi_a^{(b)}| = \chi_a^{(|b|}$ These are the only functions used in the Veblen-Schütte system that a relevant to our purposes here, though all of the other functions could b

[44] O. Veblen, 'Continuous increasing functions of finite and transfinite ordinal *Transactions of the American Mathematical Society*, Vol. 9 (1908), pp. 280–92.

[45] K. Schütte, 'Kennzeichnung von Ordnungszahlen durch rekursiv erklärte Fun tionen', *Mathematische Annalen*, Vol. 127 (1954), pp. 15–32; and in K. Schütt *Beweistheorie*, Springer-Verlag, Berlin, 1960, pp. 106–41.

imilarly introduced. (For purposes of comparison $\chi_a^{(b)}$ corresponds to the ymbol $\phi\left(\begin{smallmatrix} a & b & 1 \\ 0 & 1 & 2 \end{smallmatrix}\right)$[46], and to $2^a \cdot 3^b \cdot 5$.[47])

Now, for example, $\oplus$ is an associative operation on O_N, and we have $a \leq \chi_a^{(b)}$ and $b \leq \chi_a^{(b)}$ for any a, $b \in O_N$. Even more, many of the basic roperties of these notations and the $<$ relation between them can be ormalized and proved in elementary number theory.

All the theories to be considered in this and the next section are undertood to be formalized in the language of second order number theory. Variables s, t, ..., x, y, z range over natural numbers (type 0) and S, T, ..., X, Y, Z over sets of natural numbers (type 1). In the case of ramified analysis we hall further separate the type 1 variables into different degrees.

The basic constant, relation and function symbols for type 0 are $=$, $\bar{0}$, $'$ nd a list of all primitive recursive function symbols. $\bar{n}$ is the $(n+1)$-st umeral. The only relation used otherwise is, $t \in X$, where t is a term of ype 0. By an *arithmetic formula* $\mathfrak{A}$ we mean one in which there are no ound variables of type 1, though free variables of both types are permitted. etters $\mathfrak{A}$, $\mathfrak{B}$, $\mathfrak{C}$ will range over arithmetic formulas, $\mathfrak{F}$, $\mathfrak{G}$, $\mathfrak{H}$ over arbitrary ormulas.

The basic axioms and rules for all theories considered are:

5.5) (Ax_1) *Pure classical logic for types* 0, 1. We include here modus ponens and the rule of generalization. Occasionally, it is convenient to treat this as a Gentzen-type inferential system.

(Ax_2) *The axioms for identity.*

(Ax_3) *The standard recursion equations for each primitive recursive function.*

(Ax_4) *Complete induction.* That is, for any formula $\mathfrak{F}(x)$ [$= \mathfrak{F}(x, \ldots)$], the formula

$$\mathfrak{F}(\bar{0}) \wedge \bigwedge x[\mathfrak{F}(x) \rightarrow \mathfrak{F}(x')] \rightarrow \bigwedge x \mathfrak{F}(x)$$

is taken as an axiom.

he system based on (Ax_1)–(Ax_4) is just ordinary classical number theory xpanded to include second-order logic. We call this system (Z). The system f *classical analysis* is obtained by adding all instances of the *comprehension xiom*:

.6) (Ax_5) $\bigvee X \bigwedge x[x \in X \leftrightarrow \mathfrak{F}(x)]$,

here $\mathfrak{F}(x)$[$= \mathfrak{F}(x, \ldots)$] is any formula in which X is not free.

The more constructive systems we deal with are obtained by restricting

[46] In 'Kennzeichnung von Ordnungszahlen durch rekursiv erklärte Funktionen'.
[47] In *Beweistheorie*.

the use of (Ax_5) in various ways. For example, the system of *elementar* *analysis* is obtained by adjoining to (Z) all instances of (Ax_5) in which $\mathfrak{F}(x$ is an arithmetic formula $\mathfrak{A}(x)$.

To formalize Kreisel's hyperarithmetic comprehension rule (4.11), w need to distinguish formulas in Π- and Σ-normal form. Since we do no presume the formal use of an axiom of choice, we cannot expect to carr out the standard manipulations[48] for interchanging type 0 and type quantifiers and reducing to the normal forms $\bigwedge Y\mathfrak{A}(x,Y)$ and $\bigvee Z\mathfrak{B}(x,Z)$ By an (*essentially*) *Π-formula* $\mathfrak{P}$, we mean one which can be brought to prenex normal form in which we have *only universal quantifiers of type* 1 although quantifiers of type 0 can be arbitrary and can occur in any positio relative to these type 1 quantifiers. We define similarly the notion o (*essentially*) *Σ-formula* $\mathfrak{Q}$. Then the rule considered is:

(5.7) (RULE HCR) *Given a Π-formula* $\mathfrak{P}(x)[=\mathfrak{P}(x,\ldots)]$ *and Σ-formul* $\mathfrak{Q}(x)[=\mathfrak{Q}(x,\ldots)]$ *for which* $\mathfrak{P}(x)\leftrightarrow\mathfrak{Q}(x)$ *has been derived, infe* $\bigvee X\bigwedge x[x\in X\leftrightarrow\mathfrak{P}(x)]$.

For purposes of illustration below, we shall often only consider (*strict*) Π and Σ-formulas $\mathfrak{P}(x)=\bigwedge Y\mathfrak{A}(x,Y)$ and $\mathfrak{Q}(x)=\bigwedge Z\mathfrak{B}(x,Z)$. The system o axioms and rules given by (Ax_1)–(Ax_4) and (HCR) is denoted by (H). W delay until the next section the question of formalizing ramified analysis.

For discussing proofs of well-orderings in various systems it is convenien to introduce some abbreviations. Let $\mathfrak{L}(x,y)$ be any formula with the tw free variables x, y. We also write $x\prec y$ for $\mathfrak{L}(x,y)$ and $x\preceq y$ for $\mathfrak{L}(x,y)$ $x=y$). We write $\bigwedge z\prec x(\ldots)$ for $\bigwedge z(z\prec x\rightarrow\ldots)$, and similarly $\bigwedge z\preceq x(\ldots)$ $\bigvee z\prec x(\ldots)$, $\bigvee z\preceq x(\ldots)$. $\mathrm{Lin}_{\prec}(y)$ is the conjunction of sentences express ing that $\prec$ is a linear ordering of those x with $x\preceq y$. Then for any formul $\mathfrak{F}(x)[=\mathfrak{F}(x,\ldots)]$ we take:

(5.8) (i) $\mathrm{Prog}_x\,\mathfrak{F}(x)=\bigwedge x[x\preceq x\wedge\bigwedge y\prec x\mathfrak{F}(y)\rightarrow\mathfrak{F}(x)]$;

(ii) $I_x(\mathfrak{F}(x);z)=\mathrm{Lin}_{\prec}(z)\wedge[\mathrm{Prog}_x\,\mathfrak{F}(x)\rightarrow\bigwedge x\prec z\mathfrak{F}(x)]$;

(iii) $I(z)=\bigwedge X\, I_x(x\in X;z)$.

Thus $I(z)$ expresses that transfinite induction holds up to the segment of $\prec$ determined by z, i.e. that this segment is well-ordered by $\prec$. *Unless otherwis noted below, the relation $\prec$ to be used in the following is taken to be th natural formalization of the recursively enumerable relation $<$ taken i defining O*. We shall be especially interested in seeing for what notations $I(\bar{a})$ is provable in various systems.

Suppose given a starting system (S) of axioms and rules of inference, a

[48] S. C. Kleene, 'Hierarchies of number-theory predicates', 1955, loc. cit.

given by (Ax_1)–(Ax_4) and possibly some additional axioms and rules of inference. If, as with all the theories to be considered here, we recognize that the natural numbers provide the intended model for the variables of type 0, then we should also recognize the *ω-rule* as a correct rule of inference. This is, for any formula $\mathfrak{F}(x)[= \mathfrak{F}(x, \ldots)]$:

(5.9) *from* $\mathfrak{F}(0), \mathfrak{F}(1), \ldots, \mathfrak{F}(\bar{n}), \ldots$ *infer* $\bigwedge x \mathfrak{F}(x)$.

While recognition of the correctness of this apparently quite non-constructive rule may not help us to prove any new theorems in (S), its use in proof-theoretic studies of the limits of (S) can be quite important. This approach, which was initiated by Lorenzen, has been successfully pursued by Schütte in a number of papers, the work of which has been brought together in his book.[49]

We denote by (S^+) the system obtained from (S) by dropping the axiom of induction (Ax_4) and by adjoining the ω-rule in its place. Derivations are now infinite in length; they can be thought of as labelled well-founded infinite trees. With each derivation $\mathscr{D}$ in (S^+) is associated a certain ordinal $|\mathscr{D}|$, which we call its length. If $\mathscr{D}_n$ is a derivation of $\mathfrak{F}(\bar{n})$ for each n, and $\mathscr{D}$ is obtained from the $\mathscr{D}_n$ by applying the ω-rule to conclude $\bigwedge x \mathfrak{F}(x)$, then we take $|\mathscr{D}| = sup_n(|\mathscr{D}_n| + 1)$. Lengths are determined for the other rules of inference in a natural way. By the *cut-order* of an application of cut in a derivation $\mathscr{D}$ we mean the order of logical complexity of the principal cut-formula. By the cut-order of $\mathscr{D}$ we mean the supremum of the cut-orders in $\mathscr{D}$; this is $\leqq \omega$.

Now it turns out that in a number of cases, Gentzen-type cut-elimination arguments can be applied to derivations in (S^+). Moreover, we can often determine in these cases an ordinal κ which has the following property:

(5.10) *Given any derivation* $\mathscr{D}$ *of a formula* $\mathfrak{F}$ *in* (S^+) (*which is of bounded cut-order*), *if* $|\mathscr{D}| < \kappa$ *then we can construct a cut-free derivation* $\mathscr{D}'$ *of* $\mathfrak{F}$ *with* $|\mathscr{D}'| < \kappa$.

The ordinal κ chosen for a given system (S^+) will depend on whether we restrict attention to derivations of bounded cut-order or not. For example, in the case of classical number theory (Z^+) (or corresponding first-order versions with 'free' predicate variables), it is known that we can take $\kappa = \epsilon_0$, for the case that $\mathscr{D}$ is of bounded cut order. Since each theorem of (Z) is derivable in $< \omega \cdot 2$ steps in (Z^+) with bounded cut-order, it follows that each theorem $\mathfrak{F}$ of (Z) has a cut-free derivation $\mathscr{D}$ in (Z^+) of length $|\mathscr{D}| < \epsilon_0$.

49 K. Schütte, *Beweistheorie*, op. cit.

On the other hand, if we consider derivations $\mathscr{D}$ of arbitrary cut-order, the appropriate ordinal is $\chi_0^{(2)}$, i.e. the least critical ϵ-number.

Now we can also obtain in various cases the following:

(5.11) *if* $\mathscr{D}$ *is a cut-free derivation of* $\mathrm{I}(\bar{a})$ in (S^+) *then* $|\mathscr{D}| \geqq |a|$.

In particular, this holds for (Z^+). Hence by the preceding, if $\mathrm{I}(\bar{a})$ is provable in (Z) then $\mathrm{I}(\bar{a}) < \epsilon_0$, corresponding to the classical result of Gentzen.

We want to consider now a restriction on derivations in (S^+) related to the proviso of (4.2) and which we call a condition of *autonomy*. This is that we consider only those derivations $\mathscr{D}$ of length $\alpha = |a|$ for which we already have a shorter derivation of $\mathrm{I}(\bar{a})$. More precisely,

(5.12) *we call* γ *autonomous with respect to* (S^+) *if it is in the smallest class* A *of ordinals such that*:

(i) $\omega \in A$;
(ii) *if* $\alpha, \beta \in A$ *then* $\alpha + \beta \in A$;
(iii) *if* $\alpha \in A$ *and* $\beta \leqq \alpha$ *then* $\beta \in A$;
(iv) *if* $\mathscr{D}$ *is a derivation in* (S^+) *of* $\mathrm{I}(\bar{a})$ *where* $|a| = \alpha$ *and* $|\mathscr{D}| \in A$ *then* $\alpha \in A$.

(5.13) *We denote by* $\overline{Aut}(\mathrm{S}^+)$ *the least ordinal which is not autonomous with respect to* (S^+).

The use of conditions (5.12) (i) (ii), rather than simpler conditions, are for some technical reasons which provide greater initial freedom with the rules of (S^+).

Note that we have not restricted the cut-order of derivations $\mathscr{D}$ in (5.12) (iv). Thus, as an example, it can be seen that $\mathrm{I}(\bar{\epsilon}_0)$ is provable in (Z^+) in $\omega \cdot 2$ steps with unbounded cut-order. It can then be shown that $\overline{Aut}(\mathrm{Z}^+) = \chi_0^{(2)}$.

We wish now to consider derivations in the ordinary sense (of finite length) which make use of principles analogous to the ω-rule. Suppose (S) is a system of axioms and rules as above, and that S_a is a particular recursively enumerable set of axioms. We can formalize the notion: $\mathfrak{A}$ *is a formula provable from* S_a *by means of the axioms and rules of* (S), by means of a formula $\mathrm{Pr}_{\mathrm{S}_a}(\mathrm{u})$ of number theory. Then the *formalized* ω*-rule* for S_a consists of all instances of the following

$$\bigwedge \mathrm{x}\,\mathrm{Pr}_{\mathrm{S}_a}(\ulcorner \mathfrak{F}(\bar{\mathrm{x}}) \urcorner) \rightarrow \bigwedge \mathrm{x}\,\mathfrak{F}(\mathrm{x}), \tag{5.14}$$

where $\mathfrak{F}$ is any formula with the one free variable x. This expresses: *if for each n*, $\mathrm{S}_a \vdash \mathfrak{F}(n)$ *then for all x*, $\mathfrak{F}(x)$. In general, to deal formally with arbitrary sets S_a, we use formulas $\sigma_a(\mathrm{u})$ of number theory such that $\sigma_a(\ulcorner \mathfrak{A} \urcorner)$ expresses that $\mathfrak{A} \in \mathrm{S}_a$; it is then more appropriate to write $\mathrm{Pr}_{\sigma_a}(\mathrm{u})$ instead of $\mathrm{Pr}_{\mathrm{S}_a}(\mathrm{u})$. There are many instances of (5.14) which are not derivable rom S_a if S_a is consistent. If we adjoin all instances of the formalized ω-rule, ve get a new recursively enumerable set of axioms S'_a, which we shall also lenote by $S_{a\oplus 1}$. This suggests an iteration of this procedure into the ransfinite.

Starting from Turing's idea of *ordinal logic*, we have defined[50] the notion f *recursive progression of theories* S_a *based on the formalized* ω*-rule*. Furhermore, we obtained there an existence theorem which shows that we can ind a single recursively enumerating formula $\sigma(z;x)$ such that, if we set $_a(x) = \sigma(\bar{a};x)$ and S_a = the set of $\mathfrak{F}$ such that $\sigma_a(\ulcorner\mathfrak{F}\urcorner)$ holds for each a, hen we have the following:

5.15) (i) S_0 *consists simply of the axioms of* (S);
(ii) *for any* a, $S_{a\oplus 1}$ *consists of* S_a *together with all sentences of the form* $\bigwedge x \mathrm{Pr}_{\sigma_a}(\ulcorner\mathfrak{F}(x)\urcorner) \rightarrow \bigwedge x \mathfrak{F}(x)$;
(iii) *for any* a, $S_{3\cdot 5^a} = \bigcup S_{[a]_n}[n < \omega]$.

urthermore, it was shown[51] that we can choose $\sigma(z;x)$ so that the given rogression procedure is *verifiable*, i.e., formalized versions of (5.15) (i)–(iii) re provable in (Z). This is possible since the construction only makes use f the recursion theorem, and does not depend on O. On the other hand, ur main concern is with using the S_a just when $a \in O$. In this case we have, r example, if $3\cdot 5^a \in O$ then $S_a = \bigcup S_b[b < a]$.

Once a basic system (S) is fixed, we shall write $\vdash_a \mathfrak{U}$ instead of $S_a \vdash \mathfrak{U}$ ısing the axioms and rules of (S)), $\mathrm{Pr}(\bar{a};u)$ instead of $\mathrm{Pr}_{\sigma_a}(u)$, and in eneral, $\mathrm{Pr}(z;u)$ to express that 'u is provable in the z-th system'.

In any one system S_a, proofs are proofs in the ordinary sense of the word. ut how is one to decide which systems S_a are correct and therefore legitnate to work in? If (S) is correct under a certain interpretation in which e variables of type 0 range over the natural numbers, then so is S_a for each $\in O$. This reduces the question to determining which $a \in O$. For the esults[52] we essentially assumed that one could appeal to an 'oracle' for the nswers. However, for the constructive purposes we have in mind here, it is f the essence that we limit the notations a used by a condition of autonomy:

.16) *We call c autonomous with respect to* (*the progression based on*) (S) *if it belongs to the smallest set A such that*:
(i) $0 \in A$;
(ii) *if* $b \in A$ *and* $S_b \vdash \mathrm{I}(\bar{a})$ *then* $a \in A$.

[50] S. Feferman, 'Transfinite recursive progressions of axiomatic theories', 1962, c. cit.
[51] Ibid.
[52] Ibid.

Closure of the autonomous notations under $\oplus$ and predecessors are easily obtained for the systems considered here.

(5.17) *We denote by* $\overline{Aut}$ (S) *the least* α *such that for all c autonomous with respect to* (S) *we have* $|c| < \alpha$.

It is easily seen that the set of autonomous c is recursively enumerable.[53]

With respect to an initial system (S) which is recognized as predicatively correct, both the formalized ω-rule and the use of theorems from S_c for autonomous c should be recognized as correct. This sets the stage for the proposed explications of the notion of predicative proof.

6. *Hyperarithmetic and Ramified Analysis.* The explication via Kreisel's hyperarithmetic comprehension rule is immediate. We have already defined (H) to be given by (Ax_1)–(Ax_4) together with (HCR). We can then form the progression H_a in the manner described in Section 5. Then we propose:

(6.1) *A formula* $\mathfrak{F}$ *of* (*second order*) *analysis is predicatively provable if* $H_c \vdash \mathfrak{F}$ *for some c autonomous with respect to* (H).

We have yet to explain the formalization of ramified analysis. This is first approached by a slight transformation of the basic language. Instead of the variable symbols S, T, ..., X, Y, Z (for type 1), we now introduce disjoint collections $S^a, T^a, \ldots, X^a, Y^a, Z^a$ for each a; these are called the variables of degree a (in Schütte's terminology, of *Schicht a*). A formula in the new language can contain variables of different degrees, but every variable of type 1 must have a definite degree associated with it. We call such, a *graded* formula. (If not otherwise specified, 'formula' without qualification is taken in our previous sense.)

Given a graded formula $\mathfrak{F}$, we distinguish between the *real* and *apparent* degree of $\mathfrak{F}$, which we denote by $d(\mathfrak{F})$ and $d^*(\mathfrak{F})$ respectively. $d(\mathfrak{F})$ is the maximum of all $a \oplus 1$ such that a variable X^a occurs bound in $\mathfrak{F}$ and of all b such that a variable Y^b occurs free in $\mathfrak{F}$. $d^*(\mathfrak{F})$ is the maximum of all b such that a variable X^b occurs (free or bound) in $\mathfrak{F}$.

In contrast to the general discussion of the preceding paragraph, we do not define a system (R) of ramified logic outright, and then proceed to the infinitary system (R^+) and the progression of theories R_a. Rather, we define these systems directly according to various restrictions on degree.

The infinitary system (R_c^+) is concerned only with derivations of graded formulas $\mathfrak{F}$ with $d^*(\mathfrak{F}) \leq c$. Thus it has among its stock of variables only those $S^a, \ldots, Z^a$ with $a \leq c$. Its basic logical rules are the usual ones for the

[53] Cf. S. Feferman, 'Transfinite recursive progressions of axiomatic theories', 196[illegible] loc. cit.

ropositional connectives and numerical quantifiers, together with the -rule. For type 1 quantifiers we have the usual generalization rule, as well s the following axioms and rules:

6.2) (RC_a) (*Ramified comprehension axioms*). *For each* $a \leq c$ *and each graded formula* $\mathfrak{F}$ *with* $d(\mathfrak{F}) \leq a$ *we take the axiom*

$$\bigvee X^a \bigwedge x[x \in X^a \leftrightarrow \mathfrak{F}(x)]$$

(X^a *not free in* $\mathfrak{F}$).

5.3) (LG_a) (*Limit generalization*). *For each limit notation* $a \leq c$, *and each graded formula* $\mathfrak{F}(X^a)$ *with just* X^a *free, infer* $\mathfrak{F}(X^a)$ *from* $\mathfrak{F}(X^0), \ldots, \mathfrak{F}(X^b), \ldots$ *for all* $b < a$.

oth of these correspond directly to the intended interpretation of ramified nalysis as described in Part I—namely, to (4.1) (ii) and (iii), respectively. hese rules can be adapted to *sequenzen* form when dealing with cut-limination arguments for the (R_c^+).

The notion of autonomy relative to these infinitary systems must be odified slightly. First of all, we can use only graded statements of well-rdering:

5.4) *We take* $I^b(z) = \bigwedge X^b I_x(x \in X^b; z)$.

5.5) *We call* γ *autonomous with respect to* (R^+) *if it is in the smallest class* A *of ordinals which satisfies* (5.12) (i)–(iii) *and the following modified condition*:
(iv)′ *if* $\mathscr{D}$ *is a derivation in* (R_c^+) *of* $I^b(\bar{a})$ (*so* $b \leq c$), *and if* $|c|, |\mathscr{D}| \in A$ *then* $|a| \in A$.

5.6) *We denote by* $\overline{Aut}\,(R^+)$ *the least ordinal not autonomous with respect to* (R^+) *in the sense of* (6.5).

erivations of graded well-ordering statements $I^b(\bar{a})$ are not as special as ight seem at first sight. For, by a systematic 'lifting' process, and questions f autonomy aside, we can obtain a derivation of $I^c(\bar{a})$ for any $c \geq b$. On e other hand $I^b(\bar{a}) \to I^c(\bar{a})$ whenever $c \leq b$, so we get $I^c(\bar{a})$ for all $c \in O$.

For a ramified progression R_c we also make slight modifications in the eneral definition of Section 5. The formulas of R_c are again all $\mathfrak{F}$ with $^*(\mathfrak{F}) \leq c$. As with ordinary progressions, we have available a formula r(z; u) expressing that the formula u is provable in the z-th system R_z. For ch c, R_c has as axioms, besides the basic logical axioms, all instances of duction in formulas of apparent degree $\leq c$. It has, in addition, the mified comprehension axioms (RC_a) for all $a \leq c$. The formalized ω-rule taken in, for each $a \oplus 1 \leq c$, by including all instances of $\bigwedge x Pr(\bar{a};$

$\ulcorner \mathfrak{F}(\bar{x}) \urcorner) \rightarrow \bigwedge x \mathfrak{F}(x)$, where $d^*(\mathfrak{F}) \leq a$. Finally, we need a formal axiom corresponding to the limit generalization rule (LG_a). This is:

(6.7) $\bigwedge z < \bar{a} Pr(\bar{d}; \ulcorner \bigwedge X^z \mathfrak{F}(X^z) \urcorner) \rightarrow \bigwedge X^a \mathfrak{F}(X^a)$.

Here $\mathfrak{F}(X^a)$ is any graded formula with just X^a free, a is any limit notation $\leq c$ and d is any notation with $d \oplus 1 \leq c$. This is a combined reflection principle and formal limit generalization axiom. It is mainly in this respect, the treatment at limit notations, that the ramified progression differs from Section 5.

If we keep in mind (6.5) (iv)′, it is clear how we should also adapt our earlier notion of autonomy to that of autonomy with respect to the progression of theories R_a. Again, we introduce

(6.8) *Aut* (R) = *the least ordinal greater than* $|c|$ *for all c autonomous with respect to the* R_a.

Now we take as our second proposal:

(6.9) *A graded formula (i.e., formula of ramified analysis)* $\mathfrak{F}$ *is predicatively provable if* $R_c \vdash \mathfrak{F}$ *for some c autonomous with respect to the progression of theories* R_a.

In order to compare and support the proposals (6.1) and (6.9) we have obtained several results. The first main theorem is the following:

(6.10) THEOREM. $\overline{Aut}(H^+) = \overline{Aut}(R^+) = \overline{Aut}(H) = \overline{Aut}(R) = \Gamma_0$.

We shall try to sketch some of the ideas involved in the proof of this. The first part that is obtained most directly from what is already known in the literature is that

(6.11) $$\overline{Aut}(R^+) \leqq \Gamma_0.$$

This depends upon a cut-elimination argument for the systems (R_c^+), with the following as main lemma.

(6.12) *If* $\mathscr{D}$ *is a derivation of* $\mathfrak{F}$ *in* (R_b^+) *and if* $|b| = \beta$ and $|\mathscr{D}| = \alpha$ *then we can construct a cut free derivation* $\mathscr{D}'$ *of* $\mathfrak{F}$ *in* (R_b^+) *with* $|\mathscr{D}'| \leqq \chi_\alpha^{(\beta+1)}$

Then (6.11) follows from: $\alpha, \beta < \Gamma_0 \Rightarrow \chi_\alpha^{\beta+1} < \Gamma_0$. (6.12) is very closely related to a result of Schütte.[54] However, the system he considered there lacked the important limit generalization rule (LG_a). It can be shown to work with this modification as well. The result (6.11) has been obtained

[54] *Beweistheorie*, op. cit., p. 265.

independently by Schütte, and will appear.[55] Schütte has also established directly that $\Gamma_0 \leqq \overline{Aut}\,(R^+)$. However, we shall obtain the stronger result $\Gamma_0 \leqq Aut\,(R)$; it is fairly easy to see that

$$\overline{Aut}\,(R) \leqq \overline{Aut}\,(R^+).$$

Next we show

(6.13) $$\overline{Aut}\,(H^+) \leqq \overline{Aut}\,(R^+).$$

This is accomplished by a 'translation' of (H^+) into (R^+) which is reminiscent of the argument (4.12). Given any (ungraded) formula $\mathfrak{F}$, let $\mathfrak{F}^{(c)}$ be the result of replacing every bound or free variable X in $\mathfrak{F}$ by X^c. Then we are able to associate with every derivation $\mathscr{D}$ in (H^+) a certain collection of degrees c such that each formula $\mathfrak{F}$ appearing in $\mathscr{D}$ has $\mathfrak{F}^{(c)}$ correct in those degrees. The following then turns out to be true.

(6.14) *Suppose $\mathscr{D}$ is a derivation in (H^+) of $\mathfrak{F}$ and that $|\mathscr{D}| = |a|$. Let $c = \omega^a$. Then $\mathfrak{F}^{(c)}$ is derivable in (R_c^+) by means of a $\mathscr{D}'$ with $|\mathscr{D}'| \leqq \omega^2|a|$.*

This is sufficient to obtain (6.13).

The difficult part of the proof of (6.10) lies in getting inequalities in the other directions. Consider a simpler question: what is $\overline{Aut}\,(E)$, where (E) is the system of elementary analysis? It can be shown that $E \vdash \bigwedge x(I(z) \rightarrow I(\omega^z))$, and then $E \vdash \bigwedge z(I(\epsilon_z) \rightarrow I(\epsilon_{z\oplus 1}))$. This leads us to: $E \vdash \mathrm{Prog}_z I(\epsilon_z)$. However, we can't conclude from this in (E) that $I(z) \rightarrow I(\epsilon_z)$, since the predicate $I(\epsilon_z)$ is not elementary. However, we claim that: *for every a*, $E_{a\oplus 1} \vdash I(\bar{\epsilon}_a)$. If this is established, we have $\chi_0^{(2)} \leqq \overline{Aut}\,(E)$ (and in fact, we can get equality here).

In general, given a progression of theories S_a based on the formalized ω-rule, there is a certain class of formulas $\mathfrak{G}(z)$ for which we can show $S_{a\oplus 1} \vdash \mathfrak{G}(\bar{a})$ for an extensive set of notations $a \in O_N$, which we call the *provably conditionally progressive formulas*, abbreviated *p.c.p.* These are formulas for which we can prove in S_0:

(6.15) (i) $\mathfrak{G}(\bar{0})$

(ii) $\mathfrak{G}(z) \rightarrow \mathfrak{G}(z \oplus 1)$

(iii) *the formalization of: for any limit notation a, if* $S_{a\oplus 1} \vdash \bigwedge y \prec \bar{a}\,\mathfrak{G}(y)$ *then* $S_{a\oplus 1} \vdash \mathfrak{G}(\bar{a})$.

[55] K. Schütte, 'Eine Grenze für die Beweisbarkeit der transfiniten Induktion in der verzweigten Typenlogik', *Archiv für mathematische Logik und Grundlagenforschung*, vol. 7 (1965), pp. 45–60.

It is trivial, but of great use that

(6.16) *for any formula* $\mathfrak{F}$ *the formula* $I_t(\mathfrak{F}(t); z)$ *is p.c.p.*

We have been able to show that

(6.17) *for each* $a \in O_N$ *with* $a < \Gamma_0$ *and each p.c.p.* $\mathfrak{G}$ *we have* $S_{a\oplus 1} \vdash \mathfrak{G}(\bar{a})$.

In fact this also holds for notations for Γ_0 and beyond, but we have not been able to verify whether it holds for all $a \in O_N$. (6.17) will be applied in (6.21) below.

Returning to the question of autonomy for the progression H_a, it turns out that we need, essentially, a translation of the formalized ramified analysis to obtain a workable version of $\bigwedge z(I(z) \to I(\chi_z^{(b)}))$ for high values of b. This is, taking $c = \omega^{b\oplus 1}$, $\bigwedge z(I^c(z) \to I^c(\chi_z^{(b)}))$. Now, in order to get the required translation, we also need to be able to introduce sets Y defined by a certain transfinite recursion. We think of Y as representing a sequence of sets Y_t with $\langle x, t\rangle \in Y \leftrightarrow x \in Y_t$. We write $\langle y, s\rangle \in Y \restriction t \leftrightarrow s \prec t \wedge \langle y, s\rangle \in Y$. Then for any formula $\mathfrak{A}(x, t, X, Y)$ we put

(6.18) $\quad Rc_{\mathfrak{A}}(X, Y; z) = \bigwedge x \bigwedge t \preceq z[\langle x, t\rangle \in Y \leftrightarrow \mathfrak{A}(x, t, X, Y \restriction t)]$.

Then it can be shown that:

(6.19) *If* S *is a theory such that*

(i) $S \vdash I(\bar{b})$,

(ii) $S \vdash \bigwedge X \bigvee Y Rc_{\mathfrak{A}}(X, Y; \overline{\omega^{b\oplus 1}})$ *for any arithmetic* $\mathfrak{A}$, *and*

(iii) S *is closed under rule* (HCR), *then* $S \vdash I(\overline{\chi_0^{(b)}})$.

Given X, any solution Y of $Rc_{\mathfrak{A}}(X, Y; \bar{c})$ can be uniquely described, i.e. can be given by formulas in both Π- and Σ-form. This is the basis of the following result.

(6.20) *The formula* $\bigwedge X \bigvee Y\, Rc_{\mathfrak{A}}(X, Y; z)$ *is p.c.p. for any given arithmetic* $\mathfrak{A}$ *(relative to the progression of theories* H_a).

By piecing these various results [(6.16)–(6.20)] together, we are then able to show:

(6.21) (i) *For any formula* $\mathfrak{F}$ *and any* $a < \Gamma_0$, $H_{a\oplus 1} \vdash I_x(\mathfrak{F}(x); \bar{a})$.

(ii) *For any arithmetic formula* $\mathfrak{A}$ *and* $a < \Gamma_0$,

$$H_{a\oplus 1} \vdash \bigwedge X \bigvee Y\, Rc_{\mathfrak{A}}(X, Y; \bar{a}).$$

(iii) *For any* $b < \Gamma_0$ *and* $c = \omega^{b\oplus 1}$, $H_{c\oplus 1} \vdash I(\overline{\chi_0^{(b)}})$.

This then leads to the conclusion

(6.22) $$\Gamma_0 \leqq \overline{Aut}\,(H).$$

By formalizing the translation (6.14) of (H^+) into (R^+), we also get the
llowing

.23) THEOREM. *With any* $a < \Gamma_0$ *and* $\mathfrak{F}$ *such that* $\mathrm{H}_a \vdash \mathfrak{F}$ *we can associate* $c, d < \Gamma_0$ *such that* $\mathrm{R}_d \vdash \mathfrak{F}^{(c)}$.

follows that

.24) $$\Gamma_0 \leqq \overline{Aut}\,(\mathrm{R}).$$

'ith this, our first main theorem (6.10) is established. It also follows from 23 that $\bigcup \mathrm{H}_a(a < \Gamma_0)$ has an ω-model in which the type 1 variables range ver $\mathscr{M}_{\Gamma_0}$, i.e., the hyperarithmetical sets of order $<\Gamma_0$; in fact this is the nallest such ω-model.

As we have already mentioned, a translation of ramified analysis into the $_a$ is used in establishing (6.21) and (6.22). Even more, we can 'model' arts of ramified analysis in the H_a in such a way that any X can be taken the initial set, over which a sequence of collections $\mathscr{M}_z(\mathrm{X})$ is obtained tisfying conditions related to (4.1) (ii), (iii). This is an enumerated quence, i.e., we determine Y such that $\langle\langle x, u\rangle, z\rangle \in \mathrm{Y}$ determines the ements $\lambda x(\langle\langle x, u\rangle, z\rangle \in \mathrm{Y})$ of the z-th collection for $u = 0, 1, 2, \ldots$. (From is is follows that an axiom of choice can be determined for this collection; is will be applied below.) The translation of the ramified progression into e H_a then takes the following form.

.25) THEOREM. *With any* $c < \Gamma_0$ *and* $\mathfrak{F}$ *such that* $\mathrm{R}_c \vdash \mathfrak{F}$ *we can associate* $a, b < \Gamma_0$ *such that* $\mathrm{H}_a \vdash \bigwedge \mathrm{X}\, \mathfrak{F}_{\mathscr{M}_b(\mathrm{X})}$.

Now the Theorems (6.23) and (6.25) establish the intertranslatability of $\mathrm{H}_a(a < \Gamma_0)$ and $\bigcup \mathrm{R}_a(a < \Gamma_0)$ $(a \in O_N)$. This is the main evidence we have support of the proposals (6.1) and (6.9), for it establishes the (in essence) uivalence of the two basic approaches to the notion of predicative ovability. This leads us to identify Γ_0 as *the least well-ordering which is npredicative with respect to provability*. What is striking is the stability of is identification under the different methods of proof considered in (6.10).

An interesting application of the intertranslatability results concerns a ertain *rule of choice* which was also suggested by Kreisel[56] as being predi- atively valid:

.26) (Ch) *Suppose* $\mathfrak{Q}(x, \mathrm{Y})[= \mathfrak{Q}(x, \mathrm{Y}, \ldots)]$ *is a* Σ*-formula. If*

$$\bigwedge x \bigvee \mathrm{Y}\, \mathfrak{Q}(x, \mathrm{Y})$$

[56] G. Kreisel, 'The axiom of choice and the class of hyperarithmetic functions', 62, loc. cit.

has been derived, infer

$$\bigvee X \bigwedge x \mathfrak{Q}(x, \lambda y(\langle x, y\rangle \in X)).$$

It is easily shown that (HCR) is a derived rule from (Ch). However, we have

(6.27) THEOREM. *The theory* $H_{\Gamma_0}[=\bigcup H_a(a < \Gamma_0)]$ *is closed under* Rule (Ch)

The idea for the proof is that if we have, say $H_a \vdash \bigwedge x \bigvee Y \mathfrak{Q}(x, Y, U)$ wher $a < \Gamma_0$ then we get $c, d < \Gamma_0$ with $R_d \vdash \bigwedge x \bigvee Y^c \mathfrak{Q}^{(c)}(x, Y^c, U^c)$ by (6.23) an hence $a_1, b < \Gamma_0$ with $H_{a_1} \vdash \bigwedge U_{\mathscr{M}_b(X)} \bigwedge x \bigvee Y_{\mathscr{M}_b(X)} \mathfrak{Q}_{\mathscr{M}_b(X)}(x, Y, U)$ by (6.25) In particular, we can let X represent the collection consisting of U alon and apply the axiom of choice for the collections $\mathscr{M}_b(X)$.

Because of the use of the formalized ω-rule in defining the systems H_a, i is not *prima-facie* clear that H_{Γ_0} is a subsystem of classical analysis. This i established by the following.

(6.28) THEOREM. *Let* (IR) (*inductive-recursive analysis*) *be given by th following rules, in addition to* $(Ax)_1$–$(Ax)_4$.

(i) *Rule* (HCR).

(ii) *If* $I(\bar{a})$ *is derived, infer* $I_x(\mathfrak{F}(x);\bar{a})$ *for any formula* $\mathfrak{F}$.

(iii) *If* $I(\bar{a})$ *is derived, infer* $\bigwedge X \bigvee Y\ Rc_{\mathfrak{A}}(X, Y;\bar{a})$ *for any arithmeti formula* $\mathfrak{A}$.

Then (IR) *and* H_{Γ_0} *have the same theorems. Moreover, this holds eve if we allow* I *and* Rc *to be expressed with respect to an arbitrar arithmetic relation* $x \prec y[=\mathfrak{L}(x, y)]$ *in* (ii) *and* (iii).

The inclusion $IR \subseteq H_{\Gamma_0}$ is established by (6.21) (i), (ii). The proof in th other direction involves a formalization of the infinitary consistency proc of H_{Γ_0}. We do not have space to give more details concerning this here.

The connection of the above results and proofs with the Hilbert cor sistency programme is given by the following theorem, which makes use c arguments of Kreisel[57] and Tait.[58]

(6.29) (i) $(IR)(=H_{\Gamma_0})$ *is a conservative extension, with respect to i provable number-theoretic statements, of classical first-orde number theory when* $I_x(\mathfrak{A}(x);\bar{a})$ *is added to the latter for eac* $a < \Gamma_0$, *and each arithmetic* $\mathfrak{A}$.

(ii) *The consistency of* (IR) *can be proved in primitive recursi number theory by use of certain instances of the rule of transfini induction up to* Γ_0.

[57] G. Kreisel, 'Ordinals of ramified analysis' (abstract), *Journal of Symbolic Logi* Vol. 25 (1960), pp. 390–1.

[58] W. W. Tait, 'The ϵ-substitution method', to appear.

To our knowledge, this is the largest part of classical analysis which has been proved consistent by means of such transfinite inductions over natural well-orderings, given by notations for generalized critical numbers. In this respect, it improves (in part) a consistency result by Takeuti,[59] for a system permitting the use of definitions by transfinite recursion relative to any given well-ordering; Takeuti's result was obtained by means of his notion of *ordinal diagrams*. Other consistency proofs for more substantial parts of analysis have made use of constructive or semi-constructive notions involving well-foundedness in other ways (e.g. Spector[60] by means of *bar recursion*; this paper contains a constructive proof for certain special cases).[61]

7. *Predicativity at Higher Types; Conclusion.* We conclude with a brief discussion of predicativity at higher types. Here it is convenient to take as type 1 and higher type variables ϕ, ψ, ϑ, Φ, Ψ, etc., ranging over functions and functionals (or finite sequences of such) of arbitrary finite types. The notion of (essentially) Π- and Σ-formulas is extended in a natural way to higher types.

We first note that the rule (HCR) can be rewritten as a *functionality rule*:

(7.1) (F_1) *For* $\mathfrak{Q}(x,y)[=\mathfrak{Q}(x,y,\ldots)]$ *a Σ-formula containing only variables of types* 0, 1, *from*

$$\bigwedge x \bigvee !y\,\mathfrak{Q}(x,y)$$

infer

$$\bigvee \phi \bigwedge x\,\mathfrak{Q}(x,\phi(x)).$$

To derive the rule (HCR) from this, let $\mathfrak{P}_1$, $\mathfrak{Q}_1$ be Π- and Σ-formulas such that $\bigwedge x(\mathfrak{P}_1(x) \leftrightarrow \mathfrak{Q}_1(x))$ has been proved. We want to infer $\bigvee\phi\bigwedge x[\phi(x) = \bar{0} \leftrightarrow \mathfrak{Q}_1(x)]$. Using (F_1), is sufficient to show $\bigwedge x\bigvee !y\{\mathfrak{Q}_1(x) \wedge y = \bar{0}) \vee (\sim\mathfrak{P}_1(x) \wedge y = \bar{1})\}$. Conversely, (F_1) is easily obtained from (HCR).

This suggests the following *general functionality rule for higher types*:

(7.2) (F) *For* $\mathfrak{Q}(\phi,\vartheta)$ *a Σ-formula, from*

$$\bigwedge\psi\bigvee !\vartheta\,\mathfrak{Q}(\psi,\vartheta)$$

[59] G. Takeuti, 'On the inductive definition with quantifiers of second order', *Journal of the Mathematical Society of Japan*, Vol. 13 (1961), pp. 333–41.

[60] C. Spector, 'Provably recursive functionals of analysis: a consistency proof of analysis by an extension of principles formulated in current intuitionistic mathematics', *Recursive Function Theory. Proceedings of Symposia in Pure Mathematics* (Providence), Vol. 5 (1962), pp. 1–27.

[61] Since writing this paper, we have been informed that W. Howard has obtained a constructive consistency proof for the Σ *axiom* of choice, $\vee x \vee Y\mathfrak{Q}(x,Y) \rightarrow \vee X \wedge x\mathfrak{Q}(x,\lambda(\langle x,y\rangle \in X))$, which gives an upper bound in O_N for the provable well-orderings of this system. It is not known whether this theory is stronger than IR; we conjecture that it is.

infer

$$\bigvee\Phi\bigwedge\psi\mathfrak{Q}(\psi, \Phi(\psi)).$$

The main result is the following (derivability being under $(\mathrm{Ax})_1$–$(\mathrm{Ax})_4$, wit ordinary logic and complete induction extended to higher types):

(7.3) THEOREM. *The system with* Rule (F) *for arbitrary higher types is conservative extension of the system with* Rule (F_1) (*equivalently* (HCR).

The Rule (F) can be given a direct predicative justification which can, i effect, then be formalized to prove (7.3). We have not been able to se whether there are any further rules which should be considered predicativ at higher types. If one accepts the thesis that $H_{\Gamma_0}(=\mathrm{IR})$ embodies predi cative provability for formulas of type 1, it seems reasonable to advance th thesis that addition of Rule (F) embodies predicative provability at arbitrar higher finite types. It may be expected that results corresponding to (7.3) ca be obtained for various transfinite types (with those types which would b considered, restricted by a condition of autonomy); however we have nc pursued this matter.

The 'practical' effects of Rules (HCR) and (F) in the development c predicative analysis can be tested in terms of measure theory. First, w return to the question of l.u.b. A set of real numbers can be identified wit a collection $\mathscr{M}$ contained in the collection $\mathscr{D}$ of all lower Dedekind sectio X. We write $X \leqq Y$ to indicate that the ordering relation holds for th corresponding real numbers. It can be shown that the general statemen of existence of l.u.b. for bounded $\mathscr{M}$ is not predicatively provable. Howeve suppose that $X \in \mathscr{M} \wedge Y \leqq X \rightarrow Y \in \mathscr{M}$, and that $\mathscr{M}$ is bounded abov Then the l.u.b. S of $\mathscr{M}$ can be described by

$$x \in S \leftrightarrow \bigvee X\{X \in \mathscr{M} \wedge x \in X\},$$

$$x \in S \leftrightarrow \bigwedge X\{X \in \mathscr{D} \wedge X \notin \mathscr{M} \rightarrow x \in X\},$$

i.e., $S = \bigcup\mathscr{M} = \bigcap(\mathscr{D} - \mathscr{M})$. Thus S can be proved to exist in this case. Th argument can be used to predicatively derive the formalization of th following:

(7.4) *If a non-empty connected collection $\mathscr{M}$ of real numbers* (*i.e.,* *a interval*) *is bounded above then $\mathscr{M}$ has a least upper bound.*

Similarly for 'bounded below' and 'greatest lower bound'. Since l.u.b. an g.l.b. can be proved to be unique, when they exist, we can use (F) to intrc duce the functionals l.u.b. ($\mathscr{M}$) and g.l.b. ($\mathscr{M}$) for intervals. For particula applications, one must first verify that a given collection is predicative defined, i.e., is given by a (provably) predicatively defined functional, and we must have a proof that $\mathcal{M}$ is non-empty and connected.

We should not expect that Lebesgue outer measure, considered as a functional on collections $\mathcal{M}$ of real numbers to real numbers, is a predicatively defined functional, even if we restrict attention to measurable $\mathcal{M}$. What should be expected, on the basis of the preceding discussion and that of Section 3, is that separate measure functionals can be introduced for various levels of the Borel hierarchy. From this would follow that the classically interesting consequences of the Lebesgue theory can be developed in a predicative manner. It would be interesting to see some of the details of this worked out, as an extension of the work described in Section 3. It should be emphasized again that the result of such a development would constitute a significant and constructively meaningful part of classical analysis, rather than an *ad hoc* attempt to try to reformulate classical notions in constructive or semi-constructive terms.

There are some further directions of study which may be fruitful in connection with the questions studied here. First would be to gain greater insight into the structure of various models of H_0 and of H_{Γ_0}, in particular of the ω-models with higher type objects. Another would be to see what precise shape the formalization of predicativity would take when considered in a general set-theoretical (type-free) framework. Some steps have been taken in this direction by Wang,[62] but almost entirely within a ramified approach.

Although we strongly believe that the explications proposed in this paper for the notion of predicative provability in analysis are correct, we are not convinced that the matter has been settled conclusively by the results obtained so far. It is premature to say just what would constitute final evidence concerning this question. We expect that this will be revealed, at least in part, by further study of the theories considered here.

62 H. Wang, *A Survey of Mathematical Logic*, op. cit., Chs. XXIII–XXV.

VII

AN INTERPRETATION OF THE INTUITIONISTIC SENTENTIAL LOGIC

KURT GÖDEL

ONE can interpret[1] Heyting's sentential logic in terms of the concepts of the usual sentential logic and of the concept 'p is provable' (denoted by Bp), if one assumes for the latter the following axiom system $\mathfrak{S}$:

1. $Bp \to p$
2. $Bp \to .B(p \to q) \to Bq$
3. $Bp \to BBp$

In addition, we have to assume the axioms and rules of inference of the usual sentential logic for the concepts $\to, \sim, \cdot, \vee$, plus a new rule of inference: From A one may infer BA.

Heyting's basic concepts are to be translated in the following way:

$\neg p$	$\sim Bp$
$p \supset q$	$Bp \to Bq$
$p \vee q$	$Bp \vee Bq$
$p \wedge q$	$p \cdot q$

We could also translate equally well $\neg p$ with $B \sim Bp$, and $p \wedge q$ with $Bp \cdot Bq$. The translation of an arbitrary valid formula of Heyting's system follows from $\mathfrak{S}$, whereas the translation of $p \vee \neg p$ does not follow from $\mathfrak{S}$. In general, no formula of the form $BP \vee BQ$ is provable from $\mathfrak{S}$, unless BP or BQ is provable from $\mathfrak{S}$. Presumably, a formula of Heyting's calculus is valid if and only if its translation is provable from $\mathfrak{S}$.

The system $\mathfrak{S}$ is equivalent to Lewis's system of strict implication, if Bp is translated with Np (cf. p. 15 of this number [i.e. *Ergebnisse*, Vol. 4—Ed.]) and if Lewis's system is completed with Becker's[2] 'Zusatzaxiom' $Np \prec NNp$.

From *Ergebnisse eines mathematischen Kolloquiums*, Vol. 4 (Verlag Franz Deuticke, Vienna, 1933), pp. 39–40; translated here by J. Hintikka and L. Rossi. Printed by permission of Verlag Franz Deuticke and the author.

1 Kolmogorov (*Mathematische Zeitschrift*, Vol. 35, p. 58) has given a somewhat different interpretation of the intuitionistic sentential logic, though without giving any precise formalism.

2 'Zur Logik der Modalitäten', *Jahrbuch für Philosophie und phänomenologische Forschung*, Vol. 11 (1930), p. 497.

It can be pointed out that not all formulas provable from $\mathfrak{S}$ hold for the concept 'provable in a given formal system S'. For example, $B(Bp \to p)$ never holds for the latter, i.e. it holds for no system which includes arithmetic. For otherwise e.g. $B(0 \neq 0) \to 0 \neq 0$ and hence also $\sim B(0 \neq 0)$ were provable in S, i.e. the consistency of S were provable in S.

VIII

THE PRESENT THEORY OF TURING MACHINE COMPUTABILITY[1]

HARTLEY ROGERS, JR.

THANK you for inviting me to be present at the SIAM meetings and to speak to you this morning. Subject to the constraints of a single hour lecture, I should like to give you an informal introduction to Turing machine computability or, as I shall prefer to call it, the theory of *general effective computability*. Here, the phrase 'effective computability', as you might guess, means that we have to do with operations of the sort performable by digital computers under explicit deterministic programmes of instructions. The word 'general' means that we have to do with the theory that is obtained when we remove all limitations of either time or memory upon the action of such computers, and it means that within this theory we have to do with questions of existence or nonexistence of computer methods, rather than with matters of efficiency and good design. Such a theory, obviously, will be much simpler than a theory that takes these latter, most practical matters into account. In fact, we might suspect: (i) that such a theory would be so simple as to be uninteresting; and moreover, (ii) that it would be divested of exactly those features which could give it practical significance. I hope to convince you this morning that the first of these suspicions is wrong, and that we obtain a subject matter unusually rich, complex and intriguing. The second suspicion, I shall now admit, is largely true. I can only claim that our theory is a kind of 'asymptotic form' of more difficult, realistic theories, and that as such it has, on occasion, proved suggestive and stimulating to researchers in the practical computer field. The main defence for our theory, however, is the same as that for, let us say, such a mildly esoteric subject as the set theory that has grown out of analysis—namely, that it has a naturalness, beauty and relevance to existing mathematics that justify it in itself as an object of study.

My talk will fall into two parts, approximately reflecting two main phases

Reprinted with permission from the *Journal of the Society for Industrial and Applied Mathematics*, Vol. 7 (1959), pp. 114–30.

[1] Presented by invitation to SIAM at its summer meeting at Pennsylvania State University, August 1957.

ı the historical development of its subject. Part I will outline basic concepts nd results that were formulated and obtained in the decade prior to 1943. 'art II will suggest, in a limited way, some of the further developments and pplications that have appeared at an accelerating pace up to the present me.

PART I

In this part we look at a certain class of mathematical objects, the *recursive* *ınctions*, and at various equivalent formal definitions from which this class an be obtained. To fix our terminology, let N be the set of nonnegative ıtegers. Henceforth, the words 'number' and 'integer' shall refer to mem- ers of this set. Consider the class of all mappings of N into itself. The word unction' shall refer to members of this class. 'x', 'y', 'z', ... shall denote umbers. 'f', 'g', 'h', ... shall denote functions. The recursive functions will onstitute a certain subclass of the functions.

Since the formal definitions are too long for us to give them careful treat- ıent, we adopt a compromise. We suggest the content of these definitions ırough an informal and somewhat anthropomorphic diagram. I believe ıat all the intuitive essentials will be preserved and that you will be able to asp without ambiguity the significant results in this first phase of the ıeory.

Consider a box B inside of which we have a man L with a desk, pencils ıd paper. On one side B has two slots, marked *input* and *output*. If we write number on paper and pass it through the input slot, L takes it and begins rforming certain computations. If and when he finishes, he writes down a ımber obtained from the computation and passes it back to us through the ıtput slot. Assume further that L has with him explicit deterministic structions of finite length as to how the computation is to be done. We fer to these instructions as P. Finally, assume that the supply of paper is exhaustible, and that B can be enlarged in size so that an arbitrarily large nount of paper work can be stored in it in the course of any single com- ıtation. (Indeed, this elasticity might be needed just to store the input ımber, if that number were sufficiently large.) I think we had better assume, o, that L himself is inexhaustible, since we do not care how long it takes r an operator to appear, provided that it does eventually appear after a ıite amount of computation. We refer to the system B-L-P as M.

If there is an output for every input, M *represents* (in the obvious sense) function. It is important to keep in mind that a given function might be presentable in several different ways, that is to say, obtainable through veral different instructions P. Several examples will reinforce or clarify our picture so far. The function $[f_1(x) = 2x]$ is obviously representable. Similarly, the function $[f_2(x) =$ the xth digit in decimal expansion of $\pi]$ is representable (though, of course, we would expect that a P for f_2 would be rather longer than the simplest P for f_1). Consider next the function $[f_3(x) = 1$ if a run of *exactly* x successive 7's occurs somewhere in the decimal expansion of π; $f_3(x) = 0$ otherwise]. No one knows whether f_3 is representable, since no one knows whether finite instructions exist for computing it. In contrast to f_3 consider the function $[f_4(x) = 1$ if a run of *at leas* x successive 7's occurs somewhere in the decimal expansion of π; $f_4(x) =$ (otherwise]. After a moment's reflection, you will see that f_4 must either be the constant function $[f(x) = 1]$ or it must be a function of the forn $[f(x) = 1$ for $x \leqq k$; $f(x) = 0$ for $x > k]$ for some integer k. Whichever case occurs, instructions *exist* for computing f_4. Hence f_4 is representable, though we do not know at this time how to identify the correct P.

There remains a major area of vagueness in our definition. It centres o the question: exactly what constitutes admissible instructions P, and exactl how is the behaviour of L to depend upon P? Until this is settled, we canno treat such further questions as: is every function representable; are only a countable infinity of functions representable? It is indeed conceivable tha there might be no single satisfactory class of instructions and that an precise definition could be augmented in such a way as to represent func tions not previously representable. Several alternative ways to resolve thi vagueness were developed in the 1930's in the work of Church, Göde Kleene, Post, Turing and others. Any one of these ways can be summarize as follows. First, a finite symbolism (that is, a finite alphabet and precis rules for making arbitrarily long formulas) is given, and an admissible P i taken to be any finite sequence of formulas from this symbolism. Let us ca this the *P-symbolism*. Secondly, a finite set of fixed specifications—let us ca them *L-P specifications*—is given which stipulate how the behaviour of L i to depend upon P. We shall not go more deeply into the L-P specification except to remark that they remain constant as P is varied, that they prescrit for L certain simple digital bookkeeping operations, and that these ope ations are related to the computation and to P in an elementary symbol-a a-time way so that L can deal with formulas of arbitrary length.[2]

2 For the curious reader, we elaborate somewhat further. In all of the know approaches, the L-P specifications can be reduced to the following. A finite symbolis is specified for the computations and for the input and output numbers. A rath small set of bookkeeping operations is prescribed for L. These include operations writing down symbols, operations of moving one symbol at a time backward forward in the computation to or from symbols previously written, operations moving backward or forward in P, and terminal operations for transcribing a discharging from B a possible eventual output number. L is also endowed with a fix

In the approaches of Church and of Kleene, the P-symbolism is such that ıny admissible P consists of a sequence of simple functional equations, and he L-P specifications are such that L proceeds by using the input number n a sequence of substitution operations among the given equations. In the ıpproach of Turing, the symbolism and specifications are such that the ntire B-L-P system can be viewed as a digital computer (with the unlimited upply of paper taking the form of a tape running through the computer). Roughly, to use modern computing terms, L becomes the logical compo- ıent of the computer, and P becomes its programme. In Turing's approach, he entire system M is hence called a *Turing machine*.

Having settled on any such P-symbolism and L-P specifications, we can nswer some of the questions mentioned above. The class of functions epresentable is at most countably infinite, since the class of possible dif- erent P's is countably infinite; furthermore, not all functions are represent- ble, since the class of all functions is uncountable. The question remains, ow does the class of functions representable depend upon the apparently uite arbitrary choice of symbolism and specifications? Clearly if we start vith very limited symbolism and specifications, we may obtain no functions t all, or only a finite number. What, for instance, are the relations of the lasses generated by the Church, the Kleene and the Turing approaches? low, for instance, do these classes become larger if the symbolism and pecifications are augmented? What is the relation of these classes to func- ons which from an intuitive point of view, mathematicians generally agree) call 'effectively computable'? These matters were the subject of energetic ıvestigation in the first period of our theory. Since any precise choice of ymbolism gives a precisely defined class of functions, the investigations ook the form of detailed mathematical study. The results were, to a certain xtent, unexpected. We now summarize them as a single *basic result*.

Basic Result. First it was shown that the class of functions representable nder each of the Church, Kleene and Turing approaches is the same, and , moreover, the same as that obtained under several other suggested ap- roaches. Second, it was shown that, over certain very broad families of ılargements of the symbolism and specifications, the class remains un- ıanged. In fact, if certain very reasonable criteria are laid down for what ay constitute a symbolism and specifications, it can be demonstrated that

ite short-term memory which at any point preserves symbols written or examined certain preceding operations, and L is provided with a finite set of simple rules cording to which the bookkeeping operation next to be performed and the next state his short-term memory are uniquely determined by the contents of the short-term emory taken together with the symbol written or examined last. Given any input mber, L begins by writing that number into his computation and then examining first symbol in the first formula of P.

the class of functions obtained is always a subclass of the 'maximal' class o Church, Kleene and Turing. Third, many functions agreed to be intuitively effective were investigated, and all were shown to be members of the maximal class.

Thus, by a path that is highly noninvariant (i.e., dependent on arbitrar choices), we appear to have arrived at a natural and significant class o functions. They are called the *recursive functions*. The third part of the basic result suggests that we identify the informal intuitive notion of effectivenes with the precise concept of recursiveness. This proposal is known as *Church' Thesis*. So amply has the thesis been vindicated by detailed investigation o particular cases that, in the literature today, an author often concludes tha a particular function is recursive without giving detailed argument. Th situation is analogous to the use of formal set theoretic logic in ordinar mathematics. The investigator often omits much detail, but he must b prepared to supply it if challenged. For brevity, in the remainder of this talk we shall often go directly from intuitive effectiveness to an assertion o recursiveness. The listener should remember, however, that we have som standard choice of P-symbolism and L-P specifications in mind, and tha these give our results precise mathematical content.

There is an extension of the basic result that is useful in further work. Th listener will have noted that it is easy to think of procedures which ar intuitively effective in the sense that they lead to unique computations, bu which, through failure to terminate for some inputs, do not represent func tions. Let us use the intuitive notion *quasi-effective* to include all suc procedures together with those effective procedures that do represent func tions. By a *quasi-function* let us mean a mapping which is defined on som subset of N (possibly empty, possibly $= N$) into N. Clearly any P represent a quasi-function, though it may not, in general, represent a function. Th following is now the extension of the basic result: all statements of part one, two and three of the basic result hold, with 'quasi-function' substitute for 'function' and with 'quasi-effective' substituted for 'effective'. Thus w have a natural maximal class of *recursive quasi-functions*,[3] and, moreove an extended version of Church's Thesis; all of the comments made abov about Church's Thesis and about our use of it continue to apply verbati to this extended case.

We now turn to a closer study of the relation between P and the behaviou of M. First we observe that there is an effective procedure for enumerating one at a time, all possible instructions P of a given P-symbolism. This ca be achieved, for instance, by setting up a procedure which goes throug

[3] These are more commonly called the *partial recursive functions*.

successive *stages*, and which, in the nth stage, lists all instructions P not previously listed which consist of no more than n formulas, each of which contains no more than n symbols. Each possible P occurs exactly once somewhere in the enumerated sequence. From now on we assume that we have selected a fixed P-symbolism and fixed L-P specifications in one of the known standard approaches, and that we have selected a fixed effective enumeration of the instructions P. The *index* of P shall be the position at which it occurs in the enumeration. Observe that we can effectively: (i) find a P given its index, and (ii) find the index of any given P; in either case we simply look sufficiently far in the enumeration. If x is the index of P, we shall call the corresponding B-L-P system M_x. Next, let us observe that there is also an effective procedure for enumerating all ordered pairs of integers. For instance, we can take as the zth ordered pair, the ordered pair $\langle x,y \rangle$ where x and y are the unique solution to the equation $z = \frac{1}{2}(x^2 + 2xy + y^2 + 3x + y)$. Let us choose such an enumeration and associate with it (in the obvious way) the notations $z = \tau(x,y)$, $x = \pi_1(z)$, $y = \pi_2(z)$. τ, π_1 and π_2 are effectively computable.[4] Having made the above choices, we see the following procedure is intuitively quasi-effective: given any z, find the instruction P whose index is $\pi_1(z)$; then carry out the computation made by $M_{\pi_1(z)}$ upon the input $\pi_2(z)$; if and when an output occurs, take it and make it the final output of the whole procedure. The extended version of Church's Thesis now suggests that there is a u such that M_u represents the above quasi-function, that is, it suggests the following theorem.

THEOREM I. *There exists a u such that for any x and y,* M_u *gives the same output to input $\tau(x,y)$ as M_x gives to input y.*

The theorem can be formally proved; and an appropriate detailed construction of the P for M_u, while tedious, presents no serious difficulties. For our purpose today, we shall consider an appeal to Church's Thesis as constituting proof. A single M_u, which can be used thus to simulate any other M, was called by Turing a *universal machine*. Various forms of Theorem I were discovered in the first period of the theory.

Another problem considered early in the development of the theory was the question of whether or not there is an effective procedure for identifying those instructions P which represent functions. Church's Thesis suggests the

[4] For the interested reader, we remark here that the existence of a one-one effective mapping from $N \times N$ onto N means that *dimension* will not have a place in our theory analogous to that which it has in the theory of continuous functions on the real numbers. The study of recursive functions of several variables reduces directly to that of recursive functions of one variable.

following precise reformulation: is there a recursive function f such that $f(x) = 1$ if M_x represents a function and $f(x) = 0$ if M_x does not represent a function? The answer to this was soon shown to be negative by the following argument. If such an f exists, then, using instructions for f together with instructions for the M_u of Theorem I, we can find instructions for the function $[g(x) = 0$, if $f(x) = 0$; $g(x) = y + 1$ where y is the output of M_x with input x, if $f(x) = 1]$. Let w be an index for these instructions. Then M_w represents a function, and M_w with input w yields an output y which, by definition of w, equals $y + 1$. The result follows from this contradiction. You will observe that this argument is similar to the diagonal proof by which Cantor shows the real numbers to be uncountable. In our theory, such arguments are often called 'diagonal'. Some of our later methods—for instance, the proof of Theorem V below—can be viewed as more subtle varieties of diagonal argument. In fact, our theory has been described as a 'theory of diagonalization.' As we shall see, the theory is considerably richer than such a description might at first lead us to believe.

The above negative answer is given in a somewhat stronger form in the following theorem.

THEOREM II. *The function* $[f(z) = 1$ *if* $M_{\pi_1(z)}$ *with input* $\pi_2(z)$ *yields an output, and* $f(z) = 0$ *otherwise*$]$ *is not recursive.*

Both of the above forms follow as corollaries from Theorem V which we shall prove below. Further comments about them will be made at that point. Theorem II, in various forms, is sometimes referred to as yielding 'the recursive unsolvability of the halting problem for Turing machines' (the word 'halt' being associated with the appearance of an output).

Before leaving the first phase of our theory, let us develop a few more elementary ideas, from which we can obtain among other things, a proof for Theorem II. 'A', 'B', 'C', ... shall denote subsets of N. '$\bar{A}$' shall denote the set of numbers in N but not in A. 'W_x' shall denote the set of all outputs of M_x. We make the following definitions. A set A is *recursive* if the function $[f(x) = 1$ for x in A; $f(x) = 0$ for x in $\bar{A}]$ is recursive. This corresponds intuitively to the existence of an effective procedure for deciding whether or not any number is in A. A set A is *recursively enumerable* if either A is empty or A is the range (set of all outputs) of some recursive function. This corresponds intuitively to the existence of an effective procedure for listing all members of A. The following two theorems can be proved without difficulty.

THEOREM III. *A is recursive if and only if both A and $\bar{A}$ are recursively enumerable.*

THEOREM IV. *A is recursively enumerable if and only if there exists an x such that* $A = W_x$.

I leave their proofs to you for your amusement, with the remark that justification by Church's Thesis will be permissible, that the second is harder than the first and that it will help if you treat the cases A finite and A infinite separately.

The question now arises: can we have a set which is recursively enumerable but not recursive? To settle this and to provide, at the same time, material for later illustrations, we introduce the following slightly more sophisticated concept. A set A is *productive* if there exists a recursive function f such that for all x, if W_x is contained in A, then $f(x)$ is in A but not in W_x; f is then called a *productive function* for A. Note that since every recursively enumerable set is equal to W_x for some x, a productive set cannot be recursively enumerable. We can now prove the following.

THEOREM V. *Let* $A = \{z \mid \mathrm{M}_{\pi_1(z)}$ *with input* $\pi_2(z)$ *yields an output*$\}$. *Then A is recursively enumerable, and* $\bar{A}$ *is productive.*

PROOF. Consider the following procedure: given any z carry out the computation on z of M_u of Theorem I until an output by M_u occurs; when this happens, give output z. This is a quasi-effective procedure. The set of its outputs is the desired A. Hence, going to a P for this procedure we find an index w such that M_w has A as its set of outputs. Hence, by Theorem IV, A is recursively enumerable.

Assume now that there exists a recursive function g with the following property (*): for any z and any x, $\tau(x,x)$ appears as an output of M_z if and only if $\mathrm{M}_{g(z)}$ with input x yields an output. I claim that if such a g exists, then $\bar{A}$ is productive, with the function $f(z) = \tau(g(z), g(z))$ as productive function. f is clearly recursive. Assume (**): W_z contained in $\bar{A}$. Then $f(z)$ must be in $\bar{A}$ but not in W_z. For $f(z)$ in A implies by definition of A that $\mathrm{M}_{g(z)}$ with input $g(z)$ yields an output, which in turn implies by (*) that $f(z) = \tau(g(z), g(z))$ is an output of M_z, which in turn implies by (**) that $f(z)$ is in $\bar{A}$, a contradiction; and $f(z)$ in W_z implies by (*) that $\mathrm{M}_{g(z)}$ with input $g(z)$ yields an output, which implies by definition of A that $f(z) = \tau(g(z), g(z))$ is in A, which contradicts (**). It remains to show that a recursive function g exists with property (*). Given any fixed z, consider the following quasi-effective procedure. For any x, begin computing M_z with input 0; after a few steps, start the computation for M_z with input 1; then go back and work further on input 0; then start input 2; then work further on inputs 0 and 1; etc. In this way we can obtain and effectively list all the outputs of M_z. (This should give you a hint for proving Theorem IV.) If and when we find $\tau(x,x)$ as one of these outputs, give 0 as our final output.

Instructions $\hat{P}$ can be found for this procedure without difficulty. In fact, a rather natural version of $\hat{P}$ will employ the M_u of Theorem I for computing M_z; hence this $\hat{P}$ will itself depend effectively on the parameter z. That is to say, if we now vary z, we see that there is a recursive function g such that for any z, $g(z)$ is the index of the corresponding $\hat{P}$. Thus g has property (*) and we are through. Several corollaries follow.

COROLLARY. *There exists a set which is recursively enumerable but not recursive.* (This follows by Theorem III and the fact that a productive set cannot be recursively enumerable.)

COROLLARY. *Theorem II follows* (since A is not recursive).

COROLLARY. *There exists a productive set.*

COROLLARY. *The set* $B = \{x \mid M_x$ *represents a function*$\}$ *is not recursive.* (For if it were, we could find a procedure for settling any question of the form: does M_y with input z yield an output—in contradiction to Theorem II. We could do this by associating with any such question an M_x whose instructions tell us, for *any* input t, to give the same output as M_y gives for input z. x depends effectively on y and z. Recursiveness of B would now supply an answer.)

The final corollary gives a partial explanation for the fact that the recursive functions, natural as they are, were such a long time in being discovered. The recursive functions are embedded in the recursive quasi-functions, and the general problem of distinguishing, from an instruction, whether or not we have a function, is effectively unsolvable.

One central result of the first phase of our theory remains to be described. An important tool in later research, it is essentially a fixed-point theorem, whose use in the solution of implicit function problems is similar to the use of fixed-point theorems in analysis. It is known as the *Recursion Theorem* and is due to Kleene.

THEOREM VI. *For any recursive function f, there exists an integer y such that* $M_{f(y)}$ *and* M_y *represent the same quasi-function.*

PROOF. For a fixed z, consider the following quasi-effective procedure: given any x, compute M_z for input z; if and when any output, call it w, occurs, compute M_w for input x; give the result of this as final output. Since instructions for the above procedure, using M_u from Theorem I, can be made to depend effectively on parameter z, we have a recursive function g such that $M_{g(z)}$ gives that procedure. Now the function $f(g(z))$ is a recursive function. Let v be an index for it. Consider $y = g(v)$. Then M_y is $M_{g(v)}$ which by our construction represents the same quasi-function as M_w where w is he output of M_v with input v. But M_v with input v gives $f(g(v))$. Hence $v = f(g(v)) = f(y)$. Thus M_y represents the same quasi-function as $M_{f(y)}$, and the proof is done. Note, incidentally, that the proof gives us an effective vay of going from instructions for f to the value y. Hence we have a corol-ary.

COROLLARY. *There exists a recursive function h such that for any x, if x s an index of instructions for a recursive function f, then* $M_{f(h(x))}$ *and* $M_{h(x)}$ *epresent the same quasi-function.*

This concludes our survey of the first part of the theory. A final comment s in order. If the P-symbolism and L-P specifications are given in detail, all of the above informal proofs can be replaced by detailed constructions vithin the P-symbolism. As we have remarked before, this means that, ndependent of Church's Thesis, the theorems can be given precise content. n the first period of the theory, much work went into carrying out these onstructions in detail. In the course of this work, the natural and invariant uality of the objects concerned emerged more and more clearly. The aturalness of these objects is now so well confirmed that the above informal roofs would certainly be accepted today, at least in informal communi-ation between researchers. In fact, if you have a firm intuitive grasp of our oncept from my presentation so far, you are equipped to do research ourselves, without ever getting into the details of a particular P-symbolism. It is for this reason that I used the original anthropomorphic diagram.) You are in a position to do research much as the high school algebra student s in a position to do research in number theory. This natural, almost rimitive quality of our subject matter—it has been called one of the few *bsolute* concepts to emerge from modern work on foundations of mathe-natics—explains much of the recent enthusiasm of investigators about the ubject. It also explains the somewhat extravagant remark with which Post oncluded his excellent address to the American Mathematical Society in 944.[5] 'Indeed, if general recursive function is the formal equivalent of ffective calculability, its formulation may play a role in the history of ombinatory mathematics second only to that of the formulation of the oncept of natural number.'

PART II

To pick 1943 as an epoch in our theory is somewhat arbitrary. For want f a better dividing point, I choose it because of papers of Kleene, Mostowski

[5] E. L. Post, 'Recursively enumerable sets of positive integers and their decision roblems', *Bulletin of the American Mathematical Society*, Vol. 50 (1944), pp. 284–316.

and Post,[6] completed at about that time, which helped lead investigators into new and exciting areas, and which have served as a continuing stimulus to their efforts. I now go on to survey the further developments of the theory, most but not all occurring after 1943. For simplicity, I group them into six categories, though this grouping leads to some omission and obscures certain important interconnections. We can only give a few, necessarily simple, illustrations in each category. The names of researchers mentioned represent a rather arbitrary selection. Many equally deserving workers will be unnamed. The categories are: logic and foundations; recursive unsolvability; recursive invariance; recursive structures; recursive analysis; and intrinsic definitions of recursiveness.

In the first area—applications to *logic and foundations*—the date 1943 is of little significance. Recursive function theory has had a revolutionary impact on this area from the beginning. It has been of immeasurable value: first, in supplying a natural frame of reference in which to restate and gain new insights into work already done; and second, in suggesting fruitful ways to approach old problems and to formulate new ones. We give several brief illustrations.

Given any logical system, the possible formulas of its symbolism can be effectively enumerated in a manner analogous to that suggested above for enumerating all instructions P. Just as with the P's, this gives an effective way for going from formula to number and from number to formula. Once a fixed such enumeration is chosen, the number associated with a formula is usually called its *Gödel number*. In what follows, we shall sometimes refer to a set of formulas as if it were a set of integers. This will be an abbreviated way of referring to the set of corresponding Gödel numbers. In all such cases, it will be possible to show that the result in question does not depend upon the particular effective enumeration used to get the Gödel numbers.

As first illustration, we take the Gödel incompleteness theorem, one of the most celebrated results in modern logic. Our theory gives several suggestive and concise ways of restating parts of this theorem. We make the preliminary observation that in any of the usual systems of logic, the set of provable formulas is recursively enumerable (by a procedure of listing possible proofs analogous to our suggested procedure for listing P's); in fact, recursive enumerability of the provable formulas is now generally accepted as a necessary condition for a formalism to be called a 'logical system'. By *elementary arithmetic*, we shall mean the set of all formulas

[6] E. L. Post, ibid.; S. C. Kleene, 'Recursive Predicates and Quantifiers', *Transactio of the American Mathematical Society*, Vol. 53 (1953), pp. 41–73; A. Mostowsk 'On definable sets of positive integers', *Fundamenta Mathematica*, Vol. 34 (1947 pp. 81–112

xpressible using numerals, number variables, '+', '·', '=', parentheses, and ıantifiers over the number variables—in the usual way. We give these ·rmulas the usual interpretation over the integers. Our first partial restate- ent of the Gödel Theorem now is: *the true formulas of elementary arith- etic form a productive set.* The use of the concept of truth as distinct from ·ovability is justifiable in the sense that our whole discussion can be con- ·ived of as carried out in some general set theory in which truth can be ·fined by the well known foundational methods of Tarski. The concept of uth is avoided, however, in the further restatement: *given any methods of ·oof as strong as the usual ones (e.g., Peano's axioms) then the set of non- ovable formulas of elementary arithmetic either is empty or is productive.* he second formulation directly implies the first; I leave this to you and nit details. Note how much information the first formulation gives us: ere is an effective procedure such that given any effective enumeration of ue formulas, we can explicitly find from the instructions for that enumer- ion itself, a true formula not enumerated.

The logical system known as *lower predicate calculus* furnishes a second ample. *Church's Theorem* and the *Gödel Completeness Theorem* are two of e most basic results about this system. Church's Theorem states that the t of universally true formulas is not recursive, while the Gödel Complete- ss Theorem can be partially restated as the assertion that the same set of rmulas is recursively enumerable.

This concludes our direct treatment of logic and foundations, though, as e shall see, the other categories have substantial overlap with it.

The second topic, *recursive unsolvability*, has also been of interest from e beginning. It early became clear that our theory gives a natural and ecise sense in which certain kinds of 'problem' can be described as 'un- lvable by effective methods'; namely, we call a problem *recursively un- lvable* if it is equivalent to the problem of identifying members of a nrecursive set. Thus Church's Theorem, above, asserts that the 'problem' identifying universally true formulas of lower predicate calculus is cursively unsolvable. The corollaries to Theorem V in Part I above give nilar results about the problem of identifying instructions that represent nctions and the problem of identifying ordered pairs $\langle x,y \rangle$ such that M_x th input y yields an output.

A further result of this kind is contained in recent work of Novikov and Boone who show that the word problem for groups is recursively un- lvable. They exhibit a particular finitely generated group with finitely any relations such that, if all possible words are taken as the basic Gödel- mbered formulas, the set of words reducible to the identity is a non- cursive set.

A challenging open question in this area is Hilbert's 10th problem: th problem of deciding the existence of diophantine solutions to polynomia equations. The set of equations with solutions is clearly recursively enumer able. Is it recursive? Several investigators have conjectured not, hav carried their attempted proofs quite far, but have fallen short of a fina answer.

A number of other nonrecursive sets of specific mathematical or logica interest are known. Tarski and Robinson have developed useful method for demonstrating unsolvability in a broad class of formal theories. Sub stantial work in discovering positive solvability results for certain logica systems has also been done. Thus, while the lower predicate calculus formu lation of group theory is unsolvable, that for abelian groups is solvable While that for the integers is unsolvable, that for the real numbers is solv able. Despite Church's Theorem, large subclasses of the universally tru formulas of lower predicate calculus have been the subject of solvabilit investigations.

One of the most widely studied and satisfying areas of recent work is th third area—the area of *recursive invariance*. Before going into it, we make a observation about our previous discussion. The listener will have noted tha some of our theorems and illustrations have depended upon a number c apparently arbitrary choices—for instance Theorem V involves the choice of both an enumeration of P's and an enumeration of ordered pairs. Cor sider the collection $\mathscr{G}$ of all recursive functions which map N one-one ont itself. I leave it to you to verify that $\mathscr{G}$ forms a group under ordinary con position of functions; it is called the group of *recursive permutations*. Ou previous results now acquire a more invariant significance in the sense th they can be shown to hold under any recursive permutations of the origin arbitrary enumerations, as you can easily check.

The area of *recursive invariance* is chiefly concerned with the exploratio of recursive function theory as a part of mathematics in its own right. A Dekker has suggested, the group $\mathscr{G}$ is useful in describing its subject matte much of the theory can be described as the study of those properties of se of integers which are invariant under members of $\mathscr{G}$. The properties c recursiveness, recursive enumerability and productiveness are all invarian in this way, as you may check. We call such properties *recursively invarian* Two sets are called *isomorphic* if one is an image of the other under a men ber of $\mathscr{G}$. The isomorphism classes are hence the basic objects of our theor An interesting recent result of Myhill is the theorem that the properties recursively enumerable and $\bar{A}$ productive are a complete family of i variants; that is to say, any two sets possessing both properties are is morphic. This theorem takes on special significance in view of the fa that virtually all known unsolvability results yield sets possessing these properties. A natural question then to ask is: are any two nonrecursive, recursively enumerable sets isomorphic? Post, in his paper of 1944, shows that this is not true by constructing a recursively enumerable set A such that $\bar{A}$ contains no infinite recursively enumerable set (and hence is not productive, as you can show). The interesting problem remains, what sort of structural similarity *can* be found among the nonrecursive recursively enumerable sets? Post proposed a weaker notion of similarity for use in this study, the notion of *Turing equivalence*. Two sets are Turing equivalent if each is *reducible* to the other. A is *reducible* to B if the function $[f(x) = 1$ for x in A; $f(x) = 0$ for x in $\bar{A}]$ can be computed by an M provided that (i) L has available two lists, possibly infinite, of the members and nonmembers of B respectively; and (ii) the P symbolism and L-P specifications are appropriately modified in such a way that L can be instructed to consult his B lists in the course of his computation. Turing equivalence gives us larger equivalence classes. Until very recently, all known nonrecursive recursively enumerable sets A could be shown to possess the property; *for any recursively enumerable B, B is reducible to A*. Hence all were in the same equivalence class. The problem of whether or not this was necessarily so acquired the name *Post's Problem*. The foundational significance of the problem can be seen from our earlier comment relating provability in a logical system to recursive enumerability. The problem remained unsolved until last year (1956) when Friedberg and Muchnik each gave an ingenious construction showing that the recursively enumerable sets lie in an infinity of distinct classes.

The Turing equivalence classes are sometimes called *degrees of unsolvability*; they are partially ordered under the reducibility relation. The papers of Kleene and Mostowski helped initiate their study, and demonstrated certain natural relationships between the degree of a set and the logical complexity necessary to a formal definition of that set. Much interesting work at the present time concerns this last topic. Essentially the problem is one of relating logical complexity, as measured by type and location of quantifiers, to 'height' in the reducibility ordering as measured in some appropriate weakly constructive way. Both Kleene and Mostowski have continued to contribute important ideas to this work.

As a final illustration, let us turn for variety to a proof. We illustrate the use of Theorem VI, the fixed point result, in the area of recursive invariance. A set A is called *completely productive*, if there exists a recursive function f such that for any z, $f(z)$ lies either in the intersection of A and $\bar{W}_z$ or in the intersection of W_z and $\bar{A}$. As Dekker has commented, a completely productive set is a set which fails to be recursively enumerable, and does so in the strongly constructive sense that the 'counter-example' $f(z)$ can be effectively exhibited, given any instructions P with index z. It is immediate that a completely productive set is productive. Is the converse true? Myhill has announced a positive answer. We give a proof in the following theorem.

THEOREM VII. *If A is productive, then A is completely productive.*

PROOF. Let f be a productive function for A. Let z be fixed. Take any y. Consider the quasi-effective procedure: for any input x, search for $f(y)$ in W_z; when it appears, give output $f(y)$. The index for this procedure will depend effectively on y. Applying Theorem VI, we see that for some y, this procedure itself has index y. Turning to the corollary to Theorem VI, we see that this 'fixed point' y can be found effectively from z; hence there is a recursive function h such that this $y = h(z)$; that is, $M_{h(z)}$ has as its outputs either the empty set or the single number $f(h(z))$, according as $f(h(z))$ belongs to $\bar{W}_z$ or W_z. You may now verify that, with respect to the function $[\hat{f}(z) = f(h(z))]$, A is completely productive.

The fourth area, *recursive structures*, is concerned with the study of interrelations between recursive function theory and classical mathematics. It includes the study of recursive representations of the classical concepts, of recursive analogues of the classical concepts and of direct imposition of recursive structure upon the classical concepts. This work has been pursued with increasing intensity in recent years, and has become a fascinating part of the theory.

The classical set theory of Cantor is a good illustration, and the results obtained have obvious foundational significance. Essentially, we seek to find a model of the Cantor theory within the class of all sets of integers by using the Cantorian concepts subject to the restriction that all correspondences and mappings be recursive. The ordinal numbers were studied first. Investigators have included Church, Kleene and, more recently, Markwald and Spector. If, for instance, we consider all well-orderings of integers which, as relations (sets of ordered pairs), form recursive sets; and if we then consider the ordinals represented by these orderings, we obtain a proper segment of the classical first and second number classes which is, in many respects, a successful 'effectivized' model of the latter classes. This segment turns out to be, for instance, the largest segment for which notations exist and which is closed under limits of recursive increasing sequences of order type ω. This segment also proves to be a useful tool in measuring 'height' in the reducibility ordering of degrees of unsolvability.

A similar attack on the cardinal numbers has been begun by Dekker and Myhill. Using a deceptively straightforward modification of the Cantor definition, they have uncovered a rich theory. In their model, they have found, for instance, an intermediate class of *semi-finite* infinite cardinals whose arithmetic is much closer to that of the integers than is the arithmetic of the classical cardinals.

The approach to classical set theory is still in its early stages, and much remains to be done. For example, a satisfactory theory linking cardinal and ordinal numbers does not yet exist. On a somewhat broader front, studies are being made, by Wang, Kreisel and others, of general axiomatic set theories that are subject to certain kinds of recursive constraints. The prospect of further work in this whole region is an exciting one.

Another illustration of recursive structures is found in the work of Addison in exploring the relation of recursive invariance to parts of the point set theory developed by the Polish school of topology. A number of close analogues are found to exist between the two theories; they give promising suggestions for further results in both fields.

A further illustration is the work of Rabin on recursive structure in algebraic systems. A recursive structure is imposed on algebraic objects in a manner similar to that in which topological structure is often imposed. Various new and suggestive ideas appear. A chief value of the work is that it gives a natural way of handling notions of effectiveness that already exist, somewhat awkwardly, in the algebraic theory. For example, a short, elegant ormulation and proof can be given for Van der Waerden's theorem on the existence of a factorization algorithm for separable extensions over a field possessing such an algorithm.

The fifth area, *recursive analysis*, could quite properly be included under ecursive structures. It has, however, been the object of independent exploation for some time, and I accord it separate, though brief, mention. It begins with the attempt to impose recursive structure on the usual theory of eal numbers. A class of real numbers called the *recursive reals* is defined by equiring that each have a decimal expansion representable by a recursive unction. This class proves to be a natural one in the sense that it is invariant with respect to reasonable alternative definitions; and it can be shown to form an algebraically closed field. A function r of a real variable is called *effective*, if there exists a recursive function f such that, when x is a Gödel number for an initial segment of the decimal expression for any real number , then $f(x)$ is a Gödel number for an initial segment of the decimal expression for $r(t)$, and furthermore such that the output segments become arbitrarily long if longer and longer input segments are used. I leave it to you to show, using Theorem V, that every effective function is continuous. f you have got sufficiently into the spirit of the corollaries to Theorem V, you will not find the proof difficult. If we develop a theory that limits itself o the recursive reals as the only reals eligible for any purpose, discontinuous unctions can be made to reappear.

The sixth and last category, *intrinsic definitions of recursiveness*, is an interesting and challenging one. Few results exist in this area, however; and I must admit that including it as a separate topic is a matter of personal taste. A whole complex of problems are concerned. First of all, the listener will have noted that while various of our definitions prove to be *invariant* (with respect to the group $\mathscr{G}$), they are not *intrinsic*—that is, they depend upon *some* choice. The original definition of recursive function is a good example. It depends upon the choice of a P-symbolism and L-P specifications. Can a more intrinsic formulation be found? Some small progress toward this is made by Rogers.[7] Secondly, the more general question arises: can we somehow abstract our theory away from integers, functions, etc.? Can we find a development which will be to the present recursive function theory as abstract algebra is to the theory of numbers? (This problem, incidentally, is the rock upon which attempts completely to algebraize modern logic have foundered up to the present time; though the partial systematizations of Tarski, Halmos and others have been successful as far as they go. In attempts to find general algebraic analogues to the incompleteness theorems, there has been a residue of recursive function theory that has not been eliminable.) To thus incorporate recursive function theory into general abstract algebra, at least in part, would be a considerable achievement and would doubtless give much new insight into the theory as we know it.

This concludes Part II of my talk. In closing let me make several comments about the literature. If you are interested in looking further into the subject, the basic theory of Part I is perhaps most directly accessible in the forthcoming book of Davis,[8] which uses Turing's approach. This reference is also an excellent introduction to recent unsolvability results and contains material from other areas as well. Kleene's *Introduction to Metamathematics*[9] gives a comprehensive treatment to the basic theory and contains much valuable material in the areas of logic, foundations, and recursive invariance. Many results, however, exist only in the journals at the present time. Both the Davis and Kleene volumes have good bibliographies.

[7] H. Rogers, Jr., 'Gödel numberings of partial recursive functions', *Journal of Symbolic Logic*, Vol. 23 (1958).

[8] M. Davis, *Computability and Unsolvability*, McGraw-Hill, 1958.

[9] S. C. Kleene, *Introduction to Metamathematics*, Van Nostrand, New York, 1952.

IX

MATHEMATICAL LOGIC: WHAT HAS IT DONE FOR THE PHILOSOPHY OF MATHEMATICS?

GEORG KREISEL

MECHANISTIC THEORIES OF REASONING[1]

ı §2 (ii) and throughout §2, 3 a question turned up over and over again, ven when there was no doubt that the formulation of a traditional notion 'as natural. Was it significant? or fruitful? And this in turn leads to the oubts in §0(*d*). All these questions have to be faced when one seriously 'ants to find out about the nature of our experience. But there is one area here, roughly speaking, *the significance of a notion consists in the fact that e find it significant*, namely in the study of theory making itself. Examples f this occurred in §1: mathematicians who find the notion of predicative :t especially interesting will only use axioms which are valid for predicative :ts (even if they try to think of arbitrary sets); once we suspect this psycho- ›gical 'kink' we quickly discover a predicative formulation of their actual ıathematics. Similarly in §2(*c*), mechanistically minded mathematicians ut down only those axioms for constructive functions which are consistent ith the assumption that these functions are recursive. Personally I have no oubt of the objective significance of the notion of mechanical procedure, ıd hence of recursive function, for the analysis of reasoning; but even if ne has doubts about *that*, there is no doubt of its significance for the ıalysis of reasoning actually current.

I wish to point out here a similar significance of this notion for theories ıtside mathematics itself, in particular to the subject of *molecular biology*. nce this subject is in the process of development (and, in any case, no

From *Bertrand Russell: Philosopher of the Century*, ed. by R. Schoenman (Allen & nwin Ltd., London, and Atlantic-Little, Brown & Co., Boston, 1967), pp. 266–71. The complete essay is on pp. 201–72: see Editor's note below.

[1] [This is Section 4 of Professor Kreisel's essay. His references to other sections have en preserved: they do not seem to impair the intelligibility of the main points of his teresting discussion even when it is read apart from the rest.

One reason for reprinting here only part of his essay is that some of the omitted aterial is covered in his paper 'Informal Rigour and Completeness Proofs', part of ıich is included in the present volume. For further information, especially for recent sults, on the topics discussed by Professor Kreisel, see the Addenda to the essay inted in the Schoenman volume, and the papers by him listed in the Bibliography that volume. Ed.]

general formulation has been given) one cannot expect specific counter-examples. But there seems to be no doubt about the following points: First, it is to be a *general* schema for the explanation of biological processes, including those of the higher nervous system; second, the basic elements of the explanation (master plan) are of a discrete combinatorial kind, the fitting of shapes into one of a finite number of permitted matrices; third the complexity of biological phenomena is to be the consequence of the large number of basic objects involved and not of the complexity of the laws governing the basic objects. Certainly, the *attraction* of the subjec depends on these three features. *Combinatorial basic steps iterated a (large number of times are characteristic of recursive processes.* So, if the thre properties of (current) molecular biology are to be retained, also the stabl macroscopic properties of organisms would be expected to be recursive. So suppose there is an area of macroscopic experience which the theory i intended to cover and which (we have reason to believe) satisfies non recursive laws: then the theory is defective.

The area of experience which I wish to consider is mathematics itself The argument is related to Gödel's well known interpretation of his in completeness theorem: either there are mathematical objects external t ourselves or they are our own constructions and mind is not mechanical. I differs from his in two respects: First I do not make his assumption that, i mathematical objects are our own constructions we must be expected to b able to decide all their properties; for, except under some extravagan restrictions on what one admits as the self I do not see why one shoul expect so much more control over one's mental products than over one' bodily products—which are sometimes quite surprising. Second, I shoul like to use an abstract proof of the non-mechanical nature of mind (if I ca find such a proof) for the specific purpose of examining a particula biological theory, namely projected molecular biology satisfying the thre conditions above. For a difference of emphasis, cf. (*d*) below.

(*a*) *Generalities*

(i) As pointed out on p. 226,[3] (recursive) formalization of mathematics wa considered as *positive* evidence for a mechanistic theory of reasoning. It i remarkable how little work was done on this even in areas, such as predica

2 At least, we know this: if in a stochastic process (with a finite number of state the transition probabilities are recursive, any sequence of states with non-zero prol ability is automatically recursive: for, if the process is in the state $f(n)$ at stage n, f non-zero probability, f must be isolated, and so f is the only function which (i) dominated by some given f_0 (depending on the possible states) and (ii) satisfies recursive relation $R(f,n)$ for all n. By König's lemma, f is recursive. As long as t macroscopic processes are really stable, mutations should not be important.

3 [In Schoenman, *Bertrand Russell: Philosopher of the Century*. —Ed.]

gic, where the set of valid statements is recursively enumerable. The least ne would have to do is to show that there is something mechanical about e *actual* choice of proofs, not only about the set of results: after all, one ay ride to work on a camel or a donkey and get there; but this does not ean that a camel is a donkey.

(ii) For *negative* results the situation is easier since, if (it is accepted that) e set of valid statements is not recursively enumerable then simply *no* echanistic process generates them; it is understood that by 'valid statement' one does not here mean something that is necessarily objectively true, it the psycho-physical fact of being accepted. The following points should noted:

The *first* theory of such psycho-physical phenomena one thinks of is *n*-mechanical. Consider the behaviour of the mathematical community th respect to statements in the language of first order arithmetic. This haviour seems asymptotically stable. We certainly have no *better* theory present than this: a statement will be accepted if it is true.

The objection that such a theory could not be checked is fallacious. Of urse it could not be checked formally because it is not formulated formally the sense of §2(*a*): but the whole issue is whether reasoning is mechanistic, d so it is a *petitio principii* to require that only mechanistic theories of asoning are admitted.

A more serious objection would be this: the statistical principles we use evaluating information have generally been applied to testing mechanic theories. Since the whole subject of statistical inference is a bit like okery, analysis might show that the validity of the principles is tied to e testing of mechanistic theories; cf. (*c*).

(iii) Granted a negative result of the kind described in (ii), *the particular plication* to a biological theory proposed above makes the further assumpons:

(α) Mathematical behaviour is regarded as an integral part of the exrience to be explained, and not as some corner far removed from the incipal activities of the organism.

(β) Mathematical behaviour is to be explained in terms of the basic laws emselves; in particular, one does not assume the influence of some stract objects such as sets on the organism (by §1, these abstract objects, they exist, satisfy non-recursive laws!).

(γ) (A technical assumption to be verified mathematically.) The basic ws are such that the laws for co-operative phenomena, i.e., interaction of ganisms such as involved in mutual teaching of mathematics, are also cursive; cf. footnote 2, p. 148.

A refutation of (β) would be quite interesting! (γ) is a precise technical problem: incidentally, it would be quite interesting to look at correspondin problems in classical statistical mechanics: it is known, from continuit properties of partial differential equations, that discrete classical system have recursive behaviour; but I do not know corresponding results fo co-operative phenomena.

As to (α), if one really rejects it one accepts the division between menta and 'ordinary' biological phenomena. It is foolish to discuss (α) seriousl before the hypothetical negative result of (ii) has been established. But find the following debating point amusing: compare the place of mathe matical behaviour among biological phenomena to the place of astrono mical behaviour among mechanical phenomena; the former is far remove from ordinary life, exceptionally predictable, exceptional both in the sens that the predictions are precise, and also that they were the first to be noted since astronomical phenomena played an important part in building u physical theories, should one not expect the analogue too? (I admit I'd b equally prepared to take either side in a debate.)

(*b*) *Gödel's Incompleteness Theorem*

I do not think that it establishes the non-mechanistic character of mathe matical activity even under (α)–(γ) above without his assumption that w can decide all properties of our (mental) productions. For, what it estab lishes is the non-mechanistic character of the laws satisfied by, for instanc the natural numbers: and the theory of the behaviour of arithmeticiar mentioned in (*a*) (ii) above may well be wrong!

In fact, if the description of finitist and predicative mathematics in §2(is accepted as a correct description of the behaviour of finitist, respective predicativist mathematicians, we could mention an arithmetic problem o at least, an analytic problem which neither can decide, only he'd nev know it.

It still seems to me possible, though not probable, that the natur tendency of mathematicians to be finitist or predicativist is significant f the psycho-physical nature of reasoning.

(*c*) *Intuitionistic Mathematics* (cf. p. 250,[4] assumption (✳))

Despite the fact that, as shown in §2(*c*), existing intuitionistic axiom systen are consistent with (✳), it seems quite probable that (✳) can be refuted

[4] [In Schoenman, *Bertrand Russell: Philosopher of the Century*. The assumption is t following:

$$(\ast)\ \forall x\, \exists y\, A(x,y) \rightarrow (\exists e)\ \{e \text{ is the number of a recursion equation and } \forall x\, A[x, \{e\}(x)]\}$$

for *all* $A(x,y)$ without free variables. It says that all constructive number theore functions are recursive.—Ed.]

ie basis of evident assertions about proofs, in particular, without Gödel's ssumption just mentioned in (*b*). Note that this kind of argument is *not* ibject to the criticism of statistical principles given in (*a*) (ii) above.

At first sight such a refutation seems 'unscientific' because one appeals rectly to insights about acceptable proofs (in the 'empirical' sense of 149: proofs accepted after reflection). In other words, the argument has an *priori* character, superficially not unlike arguments about Euclidean pace. But it really would be most surprising if there were not *some* respects which it is easier to find out about ourselves than about planets and ılaxies! In other words, there should be procedures of discovery that are liable for the study of our mathematical activity even though they have iled elsewhere [cf. p. 254,[5] §3(*a*), etc.].

Note however that we have not yet refuted (✳)!

') *Mechanism and Materialism*

To avoid misunderstanding: the hypothesis that reasoning is not mechantic is by no means anti-materialist or anti-physicalist. *There is no evidence at* even present day *quantum theory is a mechanistic*, *i.e.*, *recursive theory* the sense that a recursively described system has recursive behaviour. pecifically, it is not known whether there exists a physical system with a amiltonian H such that, for instance, $\sigma(n)$ is the set of possible spins in the h energy state, $\sigma(n)$ finite for each n, and $\sigma(n)$ not a recursive function of n. t is not too hard to construct such H which are 'similar' to the Hamiltonians actual systems, i.e., correspond to laws of force other than the inverse uare law.) Naturally, the interest of such a result depends on one's trust in ie quantum theory. Note that the classical notion of a *well posed problem* n the sense of Hadamard) does not apply here. For, even if the unanalysed acroscopic data are approximate, their interpretation by means of the ıantum theory may be independent of experimental error. Just as in athematics: if we know (or assume) that $f(z)$ is analytic inside C then

$$\frac{1}{2\pi i}\int \frac{f'(z)}{f(z)}\,dz$$

integral, and approximate values of f (and hence of f') on C determine ie value exactly.

It would certainly be very interesting if the *size* of the system were concted with recursiveness. This would have a bearing on biological theories ecause biological molecules are large. If one accepts §0(*b*), one would pect something of the sort; for, physicists themselves always emphasized

5 [In Schoenman, *Bertrand Russell*: *Philosopher of the Century*. Ed.]

the non-mechanistic character of the new physics; at least, I think that i what they meant by: non-materialistic applied to fundamental particles.

It should be remarked that Wigner has considered much more profoun difficulties in molecular biological theories than those above. He shows under certain assumptions (not concerned with mechanistic features), tha mere *reproduction* of the systems involved is inconsistent with the laws o (current) quantum theory. I don't feel confident that his assumptions ar proper. In any case, he himself says that the experience of consciousnes has led him to doubt molecular biological theories. But if this is so, isn't it bit much to suppose that reproduction is intimately tied to consciousness So, higher mathematics being traditionally considered as the pinnacle o thought it seems more natural to look for a contradiction between simpl minded molecular theories and the phenomena of mathematical experience (Wigner's paper is Chap. 19 of: *Logic of Personal Knowledge*, Londo (1961), Routledge–Kegan Paul.)

Marginal comments. The whole discussion above is carried out in genera terms, or, as one says, by use of highly idealized notions. The genera reason is that the phenomena considered are complicated: to make then intelligible one has to find general (abstract) features. Also in psychologica analysis it is necessary to formulate one's ideas in general terms for the sak of objectivity: if the general laws are too close to experience, knowledge o these laws is liable to influence the experience. Finally, at least to me, th use of technically advanced machinery in analysing reasoning is encourag ing; after all, Aristotle thought about reasoning; one would like to se clearly what one has that he did not have! (It is no comfort to know tha over 2000 years have passed since his time unless one sees just *how* one ha used the experience of these 2000 years.)

X

THE METAPHYSICS OF THE CALCULUS

ABRAHAM ROBINSON

1. From the end of the seventeenth century until the middle of the ineteenth, the foundations of the Differential and Integral Calculus were matter of controversy. While most students of Mathematics are aware of is fact they tend to regard the discussions which raged during that period ntirely as arguments over technical details, proceeding from logically vague Newton) or untenable (Leibniz) ideas to the methods of Cauchy and Veierstrass which meet modern standards of rigor. However, a closer study f the history of the subject reveals that those who actually took part in this ialogue were motivated or influenced quite frequently by basic philo-ophical attitudes. To them the problem of the foundations of the Calculus as largely a philosophical question, just as the problem of the foundations f Set Theory is regarded in our time as philosophical no less than technical. hus, d'Alembert states in a passage from which I have taken the title of is address[1]:

'La théorie des limites est la base de la vraie Métaphysique du calcul ifférentiel.'

It will be my purpose today to describe and analyse the interplay of hilosophical and technical ideas during several significant phases in the evelopment of the Calculus. I shall carry out this task against the back-round of Non-standard Analysis as a viable Calculus of Infinitesimals. his will enable me to give a more precise assessment of certain historical eories than has been possible hitherto.

The basic ideas of Non-standard Analysis are sketched in the next two ections. A comprehensive development of that theory will be found in obinson.[2] The last chapter of that reference also contains a more detailed iscussion of the historical issues raised in the present talk.

2. Let R be the field of real numbers. We introduce a formal language L order to express within it statements about R. The precise scope of the

From *Problems in the Philosophy of Mathematics*, ed. Imre Lakatos (North-Holland ublishing Company, Amsterdam, 1967), pp. 28–40. Reprinted by permission of the ublishers and the author.

[1] J. Le R. d'Alembert, article 'Limite' in *Encyclopédie methodique ou par ordre de atières* (Mathématiques), 3 vols., Paris-Liège, 1784–9.

[2] A. Robinson, *Non-standard Analysis*, Studies in Logic and the Foundations of athematics, Amsterdam, 1966.

language depends on the purpose in hand. We shall suppose here that we have chosen L as a very rich language. Thus, L shall include symbols for all individual real numbers, for all sets of real numbers, for all binary, ternary, quaternary, etc., relations between real numbers, and also for all sets and relations of higher order, e.g. the set of all binary relations between real numbers. In addition L shall include the connectives of negation, disjunction and implication and also variables and quantifiers. Quantification will be permitted at all levels, but we may suppose, for the sake of familiarity that type restrictions have been imposed in the usual way. Thus, L is the language of a 'type theory of order ω'. Within it one can express all facts of Real (or of Complex) Analysis. There is no need to introduce function symbols explicitly for to every function of n variables $y = f(x_1, \ldots, x_n$ there corresponds an $n+1$-ary relation $F(x_1, \ldots, x_n, y)$ which holds if and only if $y = f(x_1, \ldots, x_n)$.

Let K be the set of all sentences in L which hold (are true) in the field of real numbers, R. It follows from standard results of Predicate Logic that there exists a proper extension $^*R$ of R which is a model of K, i.e. such that all sentences of K are true also in $^*R$. However, the statement just made is correct only if the sentences of K are interpreted in $^*R$ 'in Henkin's sense'. That is to say, when interpreting phrases such as 'for all relations' (of a certain type, universal quantification) or 'for some relation' (of a certain type, existential quantification) we take into account not the totality of all relations (or sets) of the given type but only a subset of these, the so-called *internal* or *admissible* relations (or sets). In particular, if S is a set or relation in R then there is a corresponding internal set or relation $^*S$ in $^*R$, where S and $^*S$ are denoted by the same symbol in L. However not all internal entities of $^*R$ are of this kind.

The *Non-standard model of Analysis* $^*R$ is by no means unique. However once it has been chosen, the totality of its internal entities is given with it. Thus, corresponding to the set of natural numbers N in R, there is an internal set $^*N$ in $^*R$ such that $^*N$ is a proper extension of N. And $^*N$ has 'the same' properties as N, i.e. it satisfies the same sentences of L just as $^*R$ has 'the same' properties as R. N is said to be a *Non-standard model of Arithmetic*. From now on all elements (individuals) of $^*R$ will be regarded as 'real numbers', while the particular elements of R will be said to be *standard*.

$^*R$ is a non-archimedean ordered field. Thus $^*R$ contains non-trivial infinitely small (*infinitesimal*) numbers, i.e. numbers $a \neq 0$ such that $|a| <$ for all standard positive r. (0 is counted as infinitesimal, trivially.) A number is *finite* if $|a| < r$ for some standard r, otherwise a is *infinite*. The elements of $^*N - N$ are the *infinite natural numbers*. An infinite number is greater than any finite number. If a is any finite real number then there exists a uniquely determined standard real number r, called the *standard part* of a such that $r - a$ is infinitesimal or, as we shall say also, such that r is infinitely close to a, write $r \simeq a$.

3. Let $f(x)$ be an ordinary ('standard') real-valued function of a real variable, defined for $a < x < b$, where a, b are standard real numbers, $a < b$. As we pass from R to $^*R$, $f(x)$ is extended automatically so as to be defined for all x in the open interval (a, b) *in* $^*R$. As customary in Analysis, we shall denote the extended function also by $f(x)$, but we may refer to it, by way of distinction, as '$f(x)$ in $^*R$', as opposed to the original '$f(x)$ in R'.

The properties of $f(x)$ in $^*R$ are closely linked to the properties of $f(x)$ in R by the fact that R and $^*R$ satisfy the same set of sentences, K. A single but relevant example of this interconnection is as follows.

Let $f(x)$ be defined in R, as above, and let x_0 be a standard number such that $a < x_0 < b$. Suppose that $f(x_0 + \xi) \simeq f(x_0)$, i.e. that $f(x_0 + \xi) - f(x_0)$ is infinitesimal, for all infinitesimal ξ, where $f(x)$ is now considered in $^*R$. Then we claim that for every standard $\epsilon > 0$ there exists a standard $\delta > 0$ such that $|f(x_0 + \xi) - f(x_0)| < \epsilon$ for all ξ such that $|\xi| < \delta$.—Indeed if ϵ is any standard positive real number then the statement,

'There exists an $\eta > 0$ such that for all ξ, $|\xi| < \eta$ implies $|f(x_0 + \xi) - f(x_0)| < \epsilon$', can be formulated as a sentence X within L. Thus, either X or not-X holds in R. But if not-X held in R then it would belong to K and hence, would hold also in $^*R$. Since X holds in $^*R$, by its definition, we conclude that actually X holds also in R. And any f which realizes η in R must be standard since there are no other numbers in R. This proves our assertion.

We may also prove the converse, i.e. if for every standard $\epsilon > 0$ there exists a standard $\delta > 0$ such that $|f(x_0 + \xi) - f(x_0)| < \epsilon$ for all ξ such that $|\xi| < \delta$ in R, then $f(x_0 + \xi) \simeq f(x_0)$ for all infinitesimal ξ in $^*R$. *This shows that $f(x)$ is continuous at x_0 in R if and only if $f(x_0 + \xi)$ is infinitely close to $f(x_0)$ in $^*R$.*

Similarly, it can be shown that $f(x)$ is differentiable at x_0 if and only if the ratios $(f(x_0 + \xi) - f(x_0))/\xi$ have the same standard part, d, for all infinitesimal $\xi \neq 0$, and d is then the derivative of $f(x)$ at x_0 in the ordinary sense.

For a last example, let $\{s_n\}$ be an infinite sequence of real numbers in R. On passing from R to $^*R$, $\{s_n\}$ is extended so as to be defined also for infinite natural numbers n. Let s be a standard real number. It can then be proved that s is the limit of $\{s_n\}$ in the ordinary sense, $\lim_{n \to \infty} s_n = s$ if and only if s_n is infinitely close to s (or, which is the same, if s is that standard part of s_n) for all infinite natural numbers n.

The above examples may suffice in order to give a hint how the Differential and Integral Calculus can be developed within the framework of Non-standard Analysis.

4. It appears that Newton's views concerning the foundations of the Calculus were somewhat ambiguous. He referred sometimes to infinitesimals, sometimes to moments, sometimes to limits and sometimes, and perhaps preferentially, to physical notions. But although he and his successors remained vague on the cardinal points of the subject, he did envisage the notion of the limit which, ultimately, became the cornerstone of Analysis. By contrast, Leibniz and his successors wished to base the Calculus, clearly and unambiguously, on a system which includes infinitely small quantities. This approach is crystallized in the first sentence of the '*Analyse des infiniment petits pour l'intelligence des lignes courbes*' by the Marquis de l'Hospital. We mention in passing that de l'Hospital, who was a pupil of Leibniz and John Bernoulli, acknowledged his indebtedness to his two great teachers.

De l'Hospital begins with a number of definitions and axioms. We quote in translation[3]:

'Definition I. A quantity is *variable* if it increases or decreases continuously; and, on the contrary, a quantity is *constant* if it remains the same while other quantities change. Thus, for a parabola, the ordinates and abscissae are variable quantities while the parameter is a constant quantity.'

'Definition II. The infinitely small portion by which a variable increases or decreases continuously is called its difference ...'

For *difference* read *differential.* There follows an example with reference to a diagram and a corollary in which it is stated as evident that the differential of a constant quantity is zero. Next, de l'Hospital introduces the differential notation and then goes on—

'First requirement or supposition. One requires that one may substitute for one another [*prendre indifféremment l'une pour l'autre*] two quantities which differ only by an infinitely small quantity: or (which is the same) that a quantity which is increased or decreased only by a quantity which is infinitely smaller than itself may be considered to have remained the same ...'

'Second requirement or supposition. One requires that a curve may be regarded as the totality of an infinity of straight segments, each infinitely small: or (which is the same) as a polygon with an infinite number of sides which determine by the angle at which they meet, the curvature of the curve ...'

[3] G. F. A. de l'Hospital, *Analyse des infiniment petites pour l'intelligence des lignes courbes*, Paris (1st ed. 1696), 2nd ed. 1715.

Here again we have omitted references to a diagram.

In order to appreciate the significance of these lines we have to remember hat, when they were written, mathematical axioms still were regarded, in he tradition of Euclid and Archimedes, as empirical facts from which other :mpirical facts could be obtained by deductive procedures; while the lefinitions were intended to endow the terms used in the theory with an :mpirical meaning. Thus (contrary to what a scheme of this kind would ignify in our time) de l'Hospital's formulation implies a belief in the eality of the infinitely small quantities with which it is concerned. And the ame conclusion can be drawn from the preface to the book:

'Ordinary Analysis deals only with finite quantities: this one [i.e. the Analysis of the present work] penetrates as far as infinity itself. It compares he infinitely small differences of finite quantities; it discovers the relations etween these differences; and in this way makes known the relations etween finite quantities, which are, as it were, infinite compared with the nfinitely small quantities. One may even say that this Analysis extends eyond infinity: For it does not confine itself to the infinitely small differ-nces but discovers the relations between the differences of these differ-nces, ...'

It is this robust belief in the reality of infinitely small quantities which eld sway on the continent of Europe through most of the eighteenth entury. And it is this point of view which is commonly believed to have een that of Leibniz. However, although Leibniz was indeed responsible or the technique and notation of this Calculus of Infinitesimals his ideas on ne foundations of the subject were quite different and considerably more ubtle. In fact, we know from Leibniz' correspondence that he was critical f de l'Hospital's belief in the reality of infinitesimals and even more critical f Fontenelle's emphatic affirmation of this opinion.

Leibniz' own view, as published in 1689[4] and as repeated and elaborated ubsequently in a number of letters, may by summarized as follows. While pproving of the introduction of infinitely small and infinitely large quan-ties, Leibniz did not consider them as real, like the ordinary 'real' numbers, ut thought of them as ideal or fictitious, rather like the imaginary numbers. Iowever, by virtue of a general principle of continuity, these ideal numbers ere supposed to be governed by the same laws as the ordinary numbers. Ioreover, Leibniz maintained that his procedure differed from 'the style f Archimedes' only in its language [*dans les expressions*]. And in describing he style of Archimedes', i.e. the Greek method of exhaustion, he used

[4] G. W. Leibniz, 'Tentamen de motuum coelestium causis', *Acta Eruditorum*, 1689, *'athematische Schriften* (ed. C. I. Gerhardt), Vol. 5, 1858, pp. 320–8.

the following perfectly appropriate, yet strikingly modern, phrase (translated)[5]:

'One takes quantities which are as large or as small as is necessary in order that the error be smaller than a given error [*pour que l'erreur soit moindre que l'erreur donnée*] ...'

However, Leibniz, like de l'Hospital after him, stated that two quantities may be accounted equal if they differ only by an amount which is infinitely small relative to them. And on the other hand, although he did not state this explicitly within his axiomatic framework, de l'Hospital, like Leibniz, assumed that the arithmetical laws which hold for finite quantities are equally valid for infinitesimals. It is evident, and was evident at the time, that these two assumptions cannot be accommodated simultaneously within a consistent framework. They were widely accepted nevertheless, and maintained themselves for a considerable length of time since it was found that their judicious and selective use was so very fruitful. However, Non-standard Analysis shows how a relatively slight modification of these ideas leads to a consistent theory or, at least, to a theory which is consistent relative to classical Mathematics. Thus, instead of claiming that two quantities which differ only by an infinitesimal amount, e.g., x and $x + \mathrm{d}x$, are actually equal, we find only that they are equivalent in a well-defined sense, $x + \mathrm{d}x \simeq x$ and can thus be substituted for one another in some relations but not in others. At the same time, the assertion that finite and infinitary quantities have 'the same' properties is explicated by the statement that both R and $^*R$ satisfy the set of sentences K. And if we ask, for example whether $^*R$ (like R) satisfies Archimedes' axiom then the answer depends on our interpretation of the question. If by Archimedes' axiom we mean the statement that from every positive number a we can obtain a number greater than 1 by repeated addition:

$$a + a + \ldots + a \ (n \text{ times}) > 1,$$

where n is an ordinary natural number, then $^*R$ does not satisfy the axiom But if we mean by it that for any $a > 0$ there exists a natural number n (which may be infinite) and that $n \cdot a > 1$, then Archimedes' axiom does hold in $^*R$.

5. In the view of many, including the author, the problem of the nature of infinitary notions is still of central importance in the Philosophy of

[5] G. W. Leibniz, 'Mémoire de M. G. G. Leibniz touchant son sentiment sur le calcul différentiel', *Journal de Trévoux*, 1701, *Mathematische Schriften* (ed. C. I. Gerhardt), Vol. 5, 1858, p. 350.

Mathematics. To a logical positivist, the entire argument over the reality of a mathematical structure may seem pointless but even he will have to acknowledge the historical importance of the issue. To de l'Hospital, the nfinitely small and large quantities (which were still thought of as geometrical entities) represented the actual infinite. On the other hand, Leibniz stated specifically that although he believed in the actual infinite in other spheres of Philosophy, he did not assume its existence in Mathematics. He also said that he accepted the potential (or as he put it, referring to the schoolmen, 'syncategorematic') infinite as exemplified, in his view, in the number of terms of an infinite series. To sum up, Leibniz accepted the ideal, or fictitious, infinite; accepted the potential infinite; and within Mathematics, rejected or at least dispensed with, the actual infinite.

Like the proponents of the new theory, its critics also were motivated by a combination of technical and philosophical considerations. Berkeley's 'Analyst';[6] compare Strong,[7] constitutes a brilliant attack on the logical inadequacies both of the Newtonian Theory of Fluxions and of the Leibnizian Differential Calculus. In discrediting these theories, Berkeley wished to discredit also the views of the scientists on theological matters. But beyond that, and more to the point, Berkeley's distaste for the Calculus was related to the fact that he had no place for the infinitesimals in a philosophy dominated by perception.

6. The second half of the eighteenth century saw several attempts to put the Calculus on a firm footing. However, apart from d'Alembert's affirmation of the importance of the limit concept (and, possibly, some of L. N. M. Carnot's ideas, which may have influenced Cauchy), none of these made a contribution of lasting value. Lagrange's attempt to base the entire subject on the Taylor series expansion was doomed to failure although, indirectly, it may have had a positive influence on the development of the idea of a formal power series.

It is generally believed that it was Cauchy who finally put the Calculus on rigorous foundations. And it may therefore come as a surprise to learn that infinitesimals still played a vital role in his system. I translate from Cauchy's *Cours d'Analyse*:[8]

'When speaking of the continuity of functions, I was obliged to discuss

[6] G. Berkeley, *The Analyst*, 1734, *Collected Works*, Vol. 4 (ed. A. A. Luce and T. E. Jessop), London, 1951.

[7] E. W. Strong, 'Mathematical Reasoning and its objects', *George Berkeley Lectures*, University of California Publications in Philosophy, Vol. 29, Berkeley and Los Angeles, 1957, pp. 65–88.

[8] A. Cauchy, *Cours d'Analyse de l'Ecole Royale Polytechnique*, Ie partie, Analyse algébrique, 1821 (Oeuvres complètes ser. 2, Vol. 3).

the principal properties of the infinitesimal quantities, properties which constitute the foundation of the infinitesimal calculus ...'

However, Cauchy did not regard these entities as basic but tried to derive them from the notion of a variable:

'A variable is a quantity which is thought to receive successively differen values ...'

'When the successive numerical values of a variable decrease indefinitel so as to become smaller than any given number, this variable become what is called an *infinitesimal* [*infiniment petit*] or an infinitely smal quantity.'

At the same time the *limit* of a variable (when it exists) is defined as a fixed value which is approached by the variable so as to differ from it finall as little as one pleases. It follows that a variable which becomes infinitesima has zero as limit.

Cauchy did not wish to regard the infinitesimals as numbers. And th assumption that they satisfy the same laws as the ordinary numbers, whicl had been stated explicitly by Leibniz, was rejected by Cauchy as un warranted. Moreover, Cauchy stated, on a later occasion, that while in finitesimals might legitimately be used in an argument they had no place i the final conclusion.

However, Cauchy's professed opinions in these matters notwithstanding he did in fact treat infinitesimals habitually as if they were ordinary number and satisfied the familiar rules of Arithmetic. And, as it happens, thi procedure led him to the correct result in most cases although there is famous and much discussed situation in the theory of series of functions i which he was led to the wrong conclusion. Here again, Non-standar Analysis, in spite of its different background, provides a remarkably ap propriate tool for the discussion of Cauchy's successes and failures.

For example, the fact that a function $f(x)$ is continuous at a point x_0 if th difference $f(x_0+\xi)-f(x_0)$ is infinitesimal for infinitesimal ξ, which is *theorem* of Non-standard Analysis (see Section 3 above), is also a precis explication of Cauchy's notion of continuity. On the other hand, in arrivin at the wrong conclusion that the sum of a series of continuous functions i continuous provided it exists, Cauchy used the unwarranted argument th if $\lim_{n\to\infty} s_n(x) = s(x)$ over an interval then $s_n(x_0) - s(x_0)$ is, *for all* x_0 *in th interval*, infinitesimal for infinite n. In Non-standard Analysis, it turns ou that this is true for standard (ordinary) $s_n(x)$, $s(x)$ and x_0, but not in gener for non-standard x_0, e.g. not if $x_0 = x_1 + \xi$ where x_1 is standard and ξ infinitesimal.

In order to appreciate to what extent Cauchy regarded the infinitesima s an integral part of his system, it is instructive to consider his definition f a derivative. To him, $f'(x)$, wherever it exists, is the limit of the ratio

$$\frac{\Delta y}{\Delta x}=\frac{f(x+\xi)-f(x)}{\xi}$$

here ξ is infinitesimal. In the standard modern approach the assumption at ξ is infinitesimal is completely redundant or, more precisely, meaning-ss. The fact that it was nevertheless introduced explicitly by Cauchy shows at his mental image of the situation was fundamentally different from urs. Thus, it would appear that, to his mind, a variable does not attain the nit zero directly but only after travelling through a region of infinitesimals.

We have to add that in our 'classical' framework the entire notion of a riable in Cauchy's sense, as a mathematical entity *sui generis*, has no ace. We might describe a variable, in a jocular mood, as a function which s lost its argument, while Cauchy's infinitesimals still are, to use Berkeley's mous phrase, the ghosts of departed quantities. But such carping critisicm oes not help us to understand the just recognition accorded to Cauchy's hievement, which is still thought by many to have resolved the funda-ental difficulties that had beset the Calculus previously.

If we wish to find the reasons for Cauchy's success we have to consider, ce again, both the technical-mathematical and the basic philosophical pects of the situation. Cauchy established the central position of the limit ncept for good. It is true that d'Alembert, who had emphasized the portance of this concept some decades earlier, in a sense went further an Cauchy by stating[9]:

'We say that in the *Differential* Calculus there are no infinitely small antities at all ...'

But apparently d'Alembert did not work out the consequences of his neral principles; while the vast scope and the subtlety of Cauchy's mathe-atical achievement showed to the world that his tools enabled him to go ther and deeper than his predecessors. He introduced these tools at a ne when the great achievements of the earlier and technically more primi-e method of infinitesimals had become commonplace. Thus, the momen-n which had enabled that method to disregard earlier attacks such as rkeley's was exhausted before the end of the eighteenth century and due ention was again given to its logical weaknesses (which had been there, all to see, all the time). These weaknesses had been associated throughout th the introduction of entities which were commonly regarded as denizens

J. Le R. d'Alembert, article 'Differentiel' in *Encyclopédie méthodique ou par ordre matières* (Mathématiques), 3 vols., Paris-Liège, 1784–9.

of the world of actual infinity. It now appeared that Cauchy was able t remove them from that domain and to base Analysis on the potential infinit (compare Cantor and Carruccio[10]). He did this by choosing as basic th notion of a variable which, intuitively, suggests potentiality rather tha actuality. And so it happened that a grateful public was willing to overloo the fact that, from a strictly logical point of view, the new method share some of the weaknesses of its predecessors and, indeed, introduced ne weaknesses of its own.

7. When Weierstrass (who had been anticipated to some extent b Bolzano) introduced the δ,ϵ-method about the middle of the nineteent century he maintained the limit concept in its central place. At the sam time, Weierstrass' approach is perhaps closer than Cauchy's to the Gree method of exhaustion or at least to the feature of that method which w described by Leibniz ('*pour que l'erreur soit moindre que l'erreur donnée* see section 4 above). On the issue of the actual infinite versus the potenti infinite, the δ,ϵ-method did not, as such, force its proponents into a defini position. To us, who are trained in the set-theoretic tradition, a phrase suc as 'for every positive ϵ, there exists a positive δ ...' does in fact seem t contain a clear reference to a well-defined infinite totality, i.e., the totali of positive real numbers. On the other hand, already Kronecker made clear, in his lectures, that to him the phrase meant that one could *compu* for, every *specified* positive ϵ, a positive δ with the required propert However, it was not then known that the abstract and the constructi approaches actually lead to different theories of Analysis, so that a math matician's inability to provide a procedure for computing a function who existence he has proved by abstract arguments is not necessarily due to h personal inadequacy.

At the same time it is rather natural that Set Theory should have arise as it did, from the consideration of certain problems of Analysis whi required the further clarification of basic concepts. And its creator, Geo Cantor, argued forcefully and in great detail that Set Theory deals with t actual infinite. Nevertheless, Cantor's attitude towards the theory of i finitely small quantities was entirely negative, in fact he went so far as claim that he could disprove their existence by means of Set Theory. I quc (translated from Cantor, op. cit.):

'*The fact of* [the existence of] *actually-infinitely large numbers is not*

[10] G. Cantor, *Mitteilungen zur Lehre vom Transfiniten*, 1887–8, Gesammelte Abhar lungen, ed. E. Zermelo, Berlin, 1932, pp. 378–439; E. Carruccio, 'I Fondame dell'Analisi matematica nel pensiero di Agostini Cauchy', *Bolletino dell'Uni Matematica Italiana*, Ser. 3, Vol. 12, 1957, pp. 298–307.

ason for the existence of actually-infinitely small quantities; *on the contrary, e impossibility of the latter can be proved precisely by means of the former*.'

'Nor do I think that this result can be obtained in any other way *fully* and *gorously*.'

The misguided attempt which is summed up in this quotation was con-rned not only with the past but was directed against P. du Bois-Reymond id O. Stolz who had just re-established a modest but rigorous theory of on-Archimedean systems. It may be recalled that, at the time, Cantor was ghting hard in order to obtain recognition for his own theory.

Cantor's belief in the actual existence of the infinities of Set Theory still edominates in the mathematical world today. His basic philosophy may likened to that of de l'Hospital and Fontenelle although their infinite iantities were thought to be concrete and geometrical while Cantor's finities are abstract and divorced from the physical world. Similarly, the tuitionists and other constructivists of our time may be regarded as the irs to the Aristotelian traditions of basing Mathematics on the potential finite. Finally, Leibniz' approach is akin to Hilbert's original formalism, r Leibniz, like Hilbert, regarded infinitary entities as ideal, or fictitious, lditions to concrete Mathematics. Thus, we may conclude this talk with e observation that although the very subject matter of foundational search has changed radically over the last two hundred years, there is a markable permanency in the concern with the infinite in Mathematics and the various philosophical attitudes which have been adopted towards this otion.

XI

WHAT IS ELEMENTARY GEOMETRY?

ALFRED TARSKI

IN colloquial language the term *elementary geometry* is used loosely to refe to the body of notions and theorems which, following the tradition c Euclid's *Elements*, form the subject matter of geometry courses in secondar schools. Thus the term has no well determined meaning and can be sul jected to various interpretations. If we wish to make elementary geometr a topic of metamathematical investigation and to obtain exact results (nc within, but) about this discipline, then a choice of a definite interpretatio becomes necessary. In fact, we have then to describe precisely whic sentences can be formulated in elementary geometry and which amor them can be recognized as valid; in other words, we have to determine th means of expression and proof with which the discipline is provided.

In this paper we shall primarily concern ourselves with a conception c elementary geometry which can roughly be described as follows: *we regar as elementary that part of Euclidean geometry which can be formulated ar established without the help of any set-theoretical devices.*[1]

More precisely, elementary geometry is conceived here as a theory wit standard formalization in the sense of Tarski, Mostowski, and Robinson

From *The Axiomatic Method, with Special Reference to Geometry and Physics*, e L. Henkin, P. Suppes, and A. Tarski (North-Holland Publishing Company, Amste dam, 1959), pp. 16–29. Reprinted by permission of the publishers and the author.

1 The paper was prepared for publication while the author was working on research project in the foundations of mathematics sponsored by the U.S. Natior Science Foundation.

2 One of the main purposes of this paper is to exhibit the significance of notio and methods of modern logic and metamathematics for the study of the foundatio of geometry. For logical and metamathematical notions involved in the discussi consult A. Tarski, 'Contributions to the theory of models', *Indagationes Math maticae*, Vol. 16 (1954), pp. 572–88, and Vol. 17 (1955), pp. 56–64; and A. Tars A. Mostowski, and R. M. Robinson, *Undecidable Theories*, Amsterdam, 19 xi + 98 pp. The main metamathematical result upon which the discussion is bas was established in A. Tarski, *A Decision Method for Elementary Algebra and Geomet* 2nd ed., Berkeley and Los Angeles, 1951, vi + 63 pp. For algebraic notions and resu consult B. L. Van der Waerden, *Modern Algebra*, revised English edn., New Yo Vol. 1, xii + 264 pp.

Several articles in this volume [i.e., *The Axiomatic Method with Special Referen to Geometry and Physics*: see above. Ed.] are related to the present paper in metho and results. This applies in the first place to Scott (D. Scott, 'Dimensions in elementa Euclidean geometry', pp. 53–67) and Szmielew (W. Szmielew, 'Some metamath matical problems concerning elementary hyperbolic geometry', pp. 30–52), and

: is formalized within elementary logic, i.e., first-order predicate calculus. ıll the variables $x, y, z, \ldots$ occurring in this theory are assumed to range ver elements of a fixed set; the elements are referred to as points, and the t as the space. The logical constants of the theory are (i) the sentential onnectives—the negation symbol $\neg$, the implication symbol $\rightarrow$, the disınction symbol $\vee$, and the conjunction symbol $\wedge$; (ii) the quantifiers—the niversal quantifier $\bigwedge$ and the existential quantifier $\bigvee$; and (iii) two special inary predicates—the identity symbol $=$ and the diversity symbol $\neq$. As on-logical constants (primitive symbols of the theory) we could choose ny predicates denoting certain relations among points in terms of which ll geometrical notions are known to be definable. Actually we pick two redicates for this purpose: the ternary predicate β used to denote the etweenness relation and the quaternary predicate δ used to denote the quidistant relation; the formula $\beta(xyz)$ is read *y lies between x and z* he case when y coincides with x or z not being excluded), while $\delta(xyzu)$ is ad *x is as distant from y as z is from u.*

Thus, in our formalization of elementary geometry, only points are eated as individuals and are represented by (first-order) variables. Since ementary geometry has no set-theoretical basis, its formalization does not rovide for variables of higher orders and no symbols are available to present or denote geometrical figures (point sets), classes of geometrical gures, etc. It should be clear that, nevertheless, we are able to express in ur symbolism all the results which can be found in textbooks of elementary eometry and which are formulated there in terms referring to various ecial classes of geometrical figures, such as the straight lines, the circles, e segments, the triangles, the quadrangles, and, more generally, the olygons with a fixed number of vertices, as well as to certain relations etween geometrical figures in these classes, such as congruence and simirity. This is primarily a consequence of the fact that, in each of the classes st mentioned, every geometrical figure is determined by a fixed finite ımber of points. For instance, instead of saying that a point z lies on the raight line through the points x and y, we can state that either $\beta(xyz)$ or $yzx)$ or $\beta(zxy)$ holds; instead of saying that two segments with the endoints x, y and x', y' are congruent, we simply state that $\delta(xyx'y')$.[3]

me extent also to Robinson (R. M. Robinson, 'Binary relations as primitive notions elementary geometry', pp. 68–85).

[3] In various formalizations of geometry (whether elementary or not) which are ıown from the literature, and in particular in all those which follow the lines of ilbert (D. Hilbert, *Grundlagen der Geometrie*, 8th ed., with revisions and supplements P. Bernays, Stuttgart, 1956, iii + 251 pp.), not only points but also certain geoetrical figures are treated as individuals and are represented by first-order variables; ually the only figures treated this way are straight lines, planes, and, more generally,

A sentence formulated in our symbolism is regarded as valid if it follow (semantically) from sentences adopted as axioms, i.e., if it holds in ever mathematical structure in which all the axioms hold. In the present cas by virtue of the completeness theorem for elementary logic, this amounts t saying that a sentence is valid if it is derivable from the axioms by means c some familiar rules of inference. To obtain an appropriate set of axioms, w start with an axiom system which is known to provide an adequate basis fo the whole of Euclidean geometry and contains β and δ as the only non logical constants. Usually the only non-elementary sentence in such system is the continuity axiom, which contains secondary variables $X, Y, \ldots$ ranging over arbitrary point sets (in addition to first-order variables $x, y, \ldots$ ranging over points) and also an additional logical constant, the member ship symbol $\in$ denoting the membership relation between points and poir sets. The continuity axiom can be formulated, e.g., as follows:

$$\bigwedge XY\{\bigvee z \bigwedge xy[x \in X \wedge y \in Y \rightarrow \beta(zxy)] \rightarrow \bigvee u \bigwedge xy\, [x \in X \wedge y \in Y \rightarrow \beta(xuy)]$$

We remove this axiom from the system and replace it by the infinite collec tion of all elementary continuity axioms, i.e., roughly, by all the sentence which are obtained from the non-elementary axiom if $x \in X$ is replaced b an arbitrary elementary formula in which x occurs free, and $y \in Y$ by a arbitrary elementary formula in which y occurs free. To fix the ideas, w restrict ourselves in what follows to the two-dimensional elementar geometry and quote explicitly a simple axiom system obtained in the wa just described. The system consists of twelve individual axioms, A1–A1 and the infinite collection of all elementary continuity axioms, A13.

A1 [IDENTITY AXIOM FOR BETWEENNESS].

$$\bigwedge xy[\beta(xyx) \rightarrow (x = y)]$$

linear subspaces. The set-theoretical relations of membership and inclusion, betwe a point and a special geometrical figure or between two such figures, are replaced the geometrical relation of incidence, and the symbol denoting this relation is includ in the list of primitive symbols of geometry. All other geometrical figures are treat as point sets and can be represented by second-order variables (assuming that t system of geometry discussed is provided with a set-theoretical basis). This approa has some advantages for restricted purposes of projective geometry; in fact, it facilitat the development of projective geometry by yielding a convenient formulation of t duality principle, and leads to a subsumption of this geometry under the algebra theory of lattices. In other branches of geometry an analogous procedure can hard be justified; the non-uniform treatment of geometrical figures seems to be intrinsica unnatural, obscures the logical structure of the foundations of geometry, and leads some complications in the development of this discipline (by necessitating, e.g., distinction between a straight line and the set of all points on this line).

A2 [TRANSITIVITY AXIOM FOR BETWEENNESS].

$\bigwedge xyzu[\beta(xyu) \wedge \beta(yzu) \rightarrow \beta(xyz)]$

A3 [CONNECTIVITY AXIOM FOR BETWEENNESS].

$\bigwedge xyzu[\beta(xyz) \wedge \beta(xyu) \wedge (x \neq y) \rightarrow \beta(xzu) \vee \beta(xuz)]$

A4 [REFLEXIVITY AXIOM FOR EQUIDISTANCE].

$\bigwedge xy[\delta(xyyx)]$

A5 [IDENTITY AXIOM FOR EQUIDISTANCE].

$\bigwedge xyz[\delta(xyzz) \rightarrow (x = y)]$

A6 [TRANSITIVITY AXIOM FOR EQUIDISTANCE].

$\bigwedge xyzuvw[\delta(xyzu) \wedge \delta(xyvw) \rightarrow \delta(zuvw)]$

A7 [PASCH'S AXIOM].

$\bigwedge txyzu \bigvee v\, [\beta(xtu) \wedge \beta(yuz) \rightarrow \beta(xvy) \wedge \beta(ztv)]$

A8 [EUCLID'S AXIOM].

$$\bigwedge txyzu \bigvee vw\, [\beta(xut) \wedge \beta(yuz) \wedge (x \neq u) \rightarrow \beta(xzv) \wedge \beta(xyw) \wedge \beta(vtw)]$$

A9 (FIVE-SEGMENT AXIOM).

$$\bigwedge xx'yy'zz'uu'[\delta(xyx'y') \wedge \delta(yzy'z') \wedge \delta(xux'u') \wedge \delta(yuy'u') \wedge \beta(xyz) \wedge \beta(x'y'z') \wedge (x \neq y) \rightarrow \delta(zuz'u')]$$

A10 (AXIOM OF SEGMENT CONSTRUCTION).

$\bigwedge xyuv \bigvee z\, [\beta(xyz) \wedge \delta(yzuv)]$

A11 (LOWER DIMENSION AXIOM).

$\bigvee xyz[\neg \beta(xyz) \wedge \neg \beta(yzx) \wedge \neg \beta(zxy)]$

A12 (UPPER DIMENSION AXIOM).

$$\bigwedge xyzuv[\delta(xuxv) \wedge \delta(yuyv) \wedge \delta(zuzv) \wedge (u \neq v) \rightarrow \beta(xyz) \vee \beta(yzx) \vee \beta(zxy)]$$

A13 [ELEMENTARY CONTINUITY AXIOMS]. *All sentences of the form*

$$\bigwedge vw \ldots \{\bigvee z \bigwedge xy[\phi \wedge \psi \rightarrow \beta(zxy)] \rightarrow \bigvee u \bigwedge xy[\phi \wedge \psi \rightarrow \beta(xuy)]\}$$

where ϕ *stands for any formula in which the variables* $x, v, w, \ldots,$ *but neither* y *nor* z *nor* u*, occur free, and similarly for* ψ*, with* x *and* y *interchanged.*

Elementary geometry based upon the axioms just listed will be denotec by $\mathscr{E}_2$. In Theorems 1–4 below we state fundamental metamathematica properties of this theory.[4]

First we deal with the *representation problem* for $\mathscr{E}_2$, i.e., with the problen of characterizing all models of this theory. By a model of $\mathscr{E}_2$ we understanc a system $\mathfrak{M} = \langle A, B, D\rangle$ such that (i) A is an arbitrary non-empty set, an B and D are respectively a ternary and a quaternary relation among element of A; (ii) all the axioms of $\mathscr{E}_2$ prove to hold in $\mathfrak{M}$ if all the variables ar assumed to range over elements of A, and the constants β and δ are under stood to denote the relations B and D, respectively.

The most familiar examples of models of $\mathscr{E}_2$ (and ones which can easil be handled by algorithmic methods) are certain Cartesian spaces ove ordered fields. We assume known under what conditions a syster $\mathfrak{F} = \langle F, +, \cdot, \leqslant\rangle$ (where F is a set, $+$ and $\cdot$ are binary operations under whic F is closed, and $\leqslant$ is a binary relation between elements of F) is referred t as an ordered field and how the symbols 0, $x - y$, x^2 are defined for ordere fields. An ordered field $\mathfrak{F}$ will be called Euclidean if every non-negativ element in F is a square; it is called real closed if it is Euclidean and every polynomial of an odd degree with coefficients in F has a zero in Consider the set $A_{\mathfrak{F}} = F \times F$ of all ordered couples $x = \langle x_1, x_2\rangle$ with x_1 an x_2 in F. We define the relations $B_{\mathfrak{F}}$ and $D_{\mathfrak{F}}$ among such couples by mear of the following stipulations:

$$B_{\mathfrak{F}}(xyz) \textit{ if and only if } (x_1 - y_1)\cdot(y_1 - z_2) = (x_2 - y_2)\cdot(y_1 - z_1),$$

$$0 \leqslant (x_1 - y_1)\cdot(y_1 - z_1), \textit{ and } 0 \leqslant (x_2 - y_2)\cdot(y_2 - z_2$$

$$D_{\mathfrak{F}}(xyzu) \textit{ if and only if } (x_1 - y_1)^2 + (x_2 - y_2)^2 = (z_1 - u_1)^2 + (z_2 - u_2)^2.$$

The system $\mathfrak{C}_2(\mathfrak{F}) = \langle A_{\mathfrak{F}}, B_{\mathfrak{F}}, D_{\mathfrak{F}}\rangle$ is called the (two-dimensional) Ca tesian space over $\mathfrak{F}$. If in particular we take for $\mathfrak{F}$ the ordered field $\mathfrak{R}$

4 A brief discussion of the theory $\mathscr{E}_2$ and its metamathematical properties was giv in Tarski, *A Decision Method for Elementary Algebra and Geometry*, pp. 43 ff. detailed development (based upon the results of *A Decision Method*) can be fou in Schwabhäuser (W. Schwabhäuser, 'Über die Vollständigkeit der elementar euklidischen Geometrie', *Zeitschrift für Logik und Grundlagen der Mathematik*, Vol. (1956), pp. 137–65)—where, however, the underlying system of elementary geomet differs from the one discussed in this paper in its logical structure, primitive symbo and axioms.

The axiom system for $\mathscr{E}_2$ quoted in the text above is a simplified version of t system in *A Decision Method*, pp. 55 f. The simplification consists primarily in t omission of several superfluous axioms. The proof that those superfluous axioms a actually derivable from the remaining ones was obtained by Eva Kallin, Scott Tayl and the author in connection with a course in the foundations of geometry given the author at the University of California, Berkeley, during the academic year 1956–

al numbers, we obtain the ordinary (two-dimensional) analytic space (ℜ).[5]

THEOREM 1 (REPRESENTATION THEOREM). *For* $\mathfrak{M}$ *to be a model of* $\mathscr{E}_2$ *it is cessary and sufficient that* $\mathfrak{M}$ *be isomorphic with the Cartesian space* $\mathfrak{C}_2(\mathfrak{F})$ *er some real closed field* $\mathfrak{F}$.

PROOF (in outline). It is well known that all the axioms of $\mathscr{E}_2$ hold in $\mathfrak{C}_2(\mathfrak{R})$ d that therefore $\mathfrak{C}_2(\mathfrak{R})$ is a model of $\mathscr{E}_2$. By a fundamental result in *A ecision Method*, every real closed field $\mathfrak{F}$ is elementarily equivalent with field $\mathfrak{R}$, i.e., every elementary (first-order) sentence which holds in one these two fields holds also in the other. Consequently every Cartesian ace $\mathfrak{C}_2(\mathfrak{F})$ over a real closed field $\mathfrak{F}$ is elementarily equivalent with $\mathfrak{C}_2(\mathfrak{R})$ d hence is a model of $\mathscr{E}_2$; this clearly applies to all systems $\mathfrak{M}$ isomorphic th $\mathfrak{C}_2(\mathfrak{F})$ as well.

To prove the theorem in the opposite direction, we apply methods and ults of the elementary geometrical theory of proportions, which has en developed in the literature on several occasions.[6] Consider a model $= \langle A, B, D \rangle$ of $\mathscr{E}_2$; let z and u be any two distinct points of A, and F be straight line through z and u, i.e., the set of all points x such that $B(zux)$ $B(uxz)$ or $B(xzu)$. Applying some familiar geometrical constructions, define the operations $+$ and $\cdot$ on, and the relation $\leqslant$ between, any two nts x and y in F. Thus we say that $x \leqslant y$ if either $x = y$ or else $B(xzu)$ d not $B(yxu)$ or, finally, $B(zxy)$ and not $B(xzu)$; $x + y$ is defined as the que point v in F such that $D(zxyv)$ and either $z \leqslant x$ and $y \leqslant v$ or else z and $v \leqslant y$. The definition of $x \cdot y$ is more involved; it refers to some nts outside of F and is essentially based upon the properties of parallel s. Using exclusively axioms A1–A12 we show that $\mathfrak{F} = \langle F, +, \cdot, \leqslant \rangle$ is an ered field; with the help of A13 we arrive at the conclusion that $\mathfrak{F}$ is ually a real closed field. By considering a straight line G perpendicular F at the point z, we introduce a rectangular coordinate system in $\mathfrak{M}$ and establish a one-to-one correspondence between points $x, y, \ldots$ in A and ered couples of their coordinates $\bar{x} = \langle x_1, x_2 \rangle$, $\bar{y} = \langle y_1, y_2 \rangle$, ... in $F \times F$.

All the results in this paper extend (with obvious changes) to the n-dimensional for any positive integer n. To obtain an axiom system for $\mathscr{E}_n$ we have to modify two dimension axioms, A11 and A12, leaving the remaining axioms unchanged; result in Scott's 'Dimension in elementary Euclidean geometry', A11 and A12 be replaced by any sentence formulated in the symbolism of $\mathscr{E}_n$ which holds in ordinary n-dimensional analytic space but not in any m-dimensional analytic e for $m \neq n$. In constructing algebraic models for one-dimensional geometries we ordered abelian groups instead of ordered fields.

See, e.g., Hilbert, *Grundlagen der Geometrie*, pp. 51 ff.

With the help of the Pythagorean theorem (which proves to be valid in $\mathscr{E}$
we show that the formula

$$D(xyst)$$

holds for any given points $x, y, \dots$ in A if and only if the formula

$$D_{\mathfrak{F}}(\bar{x}\bar{y}\bar{s}\bar{t})$$

holds for the correlated couples of coordinates $\bar{x} = \langle x_1, x_2 \rangle, \bar{y} = \langle y_1, y_2 \rangle,$ ·
in $F \times F$, i.e., if

$$(x_1 - y_1)^2 + (x_2 - y_2{}^2 = (s_1 - t_1)^2 + (s_2 - t_2)^2;$$

an analogous conclusion is obtained for $B(xys)$. Consequently, the syster
$\mathfrak{M}$ and $\mathfrak{C}_2(\mathfrak{F})$ are isomorphic, which completes the proof.

We turn to the *completeness problem for* $\mathscr{E}_2$. A theory is called compl
if every sentence σ (formulated in the symbolism of the theory) holds eith
in every model of this theory or in no such model. For theories with standa
formalization this definition can be put in several other equivalent forn
we can say, e.g., that a theory is complete if, for every sentence σ, either
or $\neg\sigma$ is valid, or if any two models of the theory are elementarily equivaler
A theory is called consistent if it has at least one model; here, again, sever
equivalent formulations are known. If there is a model $\mathfrak{M}$ such that
sentence holds in $\mathfrak{M}$ if and only if it is valid in the given theory, then t
theory is clearly both complete and consistent, and conversely. The soluti
of the completeness problem for $\mathscr{E}_2$ is given in the following

THEOREM 2 (COMPLETENESS THEOREM). (i) *A sentence formulated in* $\mathscr{E}_2$
valid if and only if it holds in $\mathfrak{C}_2(\mathfrak{R})$;

(ii) *the theory* $\mathscr{E}_2$ *is complete* (*and consistent*).

Part (i) of this theorem follows from Theorem 1 and from a fundamen
result in *A Decision Method* which was applied in the proof of Theorem
(ii) is an immediate consequence of (i).

The next problem which will be discussed here is the *decision problem*
$\mathscr{E}_2$. It is the problem of the existence of a mechanical method which enab
us in each particular case to decide whether or not a given sentence forn
lated in $\mathscr{E}_2$ is valid. The solution of this problem is again positive:

THEOREM 3 (DECISION THEOREM). *The theory* $\mathscr{E}_2$ *is decidable.*

In fact, $\mathscr{E}_2$ is complete by Theorem 2 and is axiomatizable by its v
description (i.e., it has an axiom system such that we can always dec
whether a given sentence is an axiom). It is known, however, that ev
complete and axiomatizable theory with standard formalization is dec

able,[7] and therefore $\mathscr{E}_2$ is decidable. By analysing the discussion in *A Decision Method* we can actually obtain a decision method for $\mathscr{E}_2$.

The last metamathematical problem to be discussed for $\mathscr{E}_2$ is the *problem of finite axiomatizability*. From the description of $\mathscr{E}_2$ we see that this theory has an axiom system consisting of finitely many individual axioms and of an infinite collection of axioms falling under a single axiom schema. This axiom schema (which is the symbolic expression occurring in A13) can be slightly modified so as to form a single sentence in the system of predicate calculus with free variable first-order predicates, and all the particular axioms of the infinite collection can be obtained from this sentence by substitution. We briefly describe the whole situation by saying that the theory $\mathscr{E}_2$ is 'almost finitely axiomatizable', and now we ask the question whether $\mathscr{E}_2$ is finitely axiomatizable in the strict sense, i.e., whether the original axiom system can be replaced by an equivalent finite system of sentences formulated in $\mathscr{E}_2$. The answer is negative:

THEOREM 4 (NON-FINITIZABILITY THEOREM). *The theory $\mathscr{E}_2$ is not finitely axiomatizable.*

PROOF (in outline). From the proof of Theorem 1 it is seen that the infinite collection of axioms A13 can be equivalently replaced by an infinite sequence of sentences $S_0, \ldots, S_n, \ldots$; S_0 states that the ordered field $\mathfrak{F}$ constructed in the proof of Theorem 1 is Euclidean, and S_n for $n > 0$ expresses the fact that in this field every polynomial of degree $2n + 1$ has a zero. For every prime number p we can easily construct an ordered field $\mathfrak{F}_p$ in which every polynomial of an odd degree $2n + 1 < p$ has a zero while some polynomial of degree p has no zero; consequently, if $2m + 1 = p$ is a prime, then all the axioms A1–A12 and S_n with $n < m$ hold in $\mathfrak{C}_2(\mathfrak{F}_p)$ while S_m does not hold. This implies immediately that the infinite axiom system A1, …, A12, $S_0, \ldots, S_n, \ldots$ has no finite subsystem from which all the axioms of the system follow. Hence by a simple argument we conclude that, more generally, there is no finite axiom system which is equivalent with the original axiom for system for $\mathscr{E}_2$.

From the proof just outlined we see that $\mathscr{E}_2$ can be based upon an axiom system A1, …, A12, $S_0, \ldots, S_n, \ldots$ in which (as opposed to the original axiom system) each axiom can be put in the form of either a universal sentence or an existential sentence or a universal-existential sentence; i.e., each axiom is either of the form

$$\bigwedge xy \ldots (\phi)$$

[7] Cf. Tarski, Mostowski, and Robinson, *Undecidable Theories*, p. 14.

or else of the form

$$\bigvee uv \ldots (\phi)$$

or, finally, of the form

$$\bigwedge xy \ldots \bigvee uv \ldots (\phi)$$

where ϕ is a formula without quantifiers. A rather obvious consequence of this structural property of the axioms is the fact that the union of a chain (or of a directed family) of models of $\mathscr{E}_2$ is again a model of $\mathscr{E}_2$. This consequence can also be derived directly from the proof of Theorem 1.

The conception of elementary geometry with which we have been concerned so far is certainly not the only feasible one. In what follows we shall discuss briefly two other possible interpretations of the term 'elementary geometry'; they will be embodied in two different formalized theories, $\mathscr{E}_2'$ and $\mathscr{E}_2''$.

The theory $\mathscr{E}_2'$ is obtained by supplementing the logical base of $\mathscr{E}_2$ with a small fragment of set theory. Specifically, we include in the symbolism of $\mathscr{E}_2'$ new variables X, Y, ... assumed to range over arbitrary finite sets of points (or, what in this case amounts essentially to the same, over arbitrary finite sequences of points); we also include a new logical constant, the membership symbol $\in$, to denote the membership relation between points and finite point sets. As axioms for $\mathscr{E}_2'$ we again choose A1–A13; it should be noticed, however, that the collection of axiom A13 is now more comprehensive than in the case of $\mathscr{E}_2$ since ϕ and ψ stand for arbitrary formulas constructed in the symbolism of $\mathscr{E}_2'$. In consequence the theory $\mathscr{E}_2'$ considerably exceeds $\mathscr{E}_2$ in means of expression and power. In $\mathscr{E}_2'$ we can formulate and study various notions which are traditionally discussed in textbooks of elementary geometry but which cannot be expressed in $\mathscr{E}_2$; e.g., the notions of a polygon with arbitrarily many vertices, and of the circumference and the area of a circle.

As regards metamathematical problems which have been discussed and solved for $\mathscr{E}_2$ in Theorems 1–4, three of them—the problems of representation, completeness, and finite axiomatizability—are still open when referred to $\mathscr{E}_2'$. In particular, we do not know any simple characterization of all models of $\mathscr{E}_2'$, nor do we know whether any two such models are equivalent with respect to all sentences formulated in $\mathscr{E}_2$. (When speaking of models of $\mathscr{E}_2'$ we mean exclusively the so-called standard models; i.e., when deciding whether a sentence σ formulated in $\mathscr{E}_2'$ holds in a given model, we assume that the variables x, y, ... occurring in σ range over all elements of a set, the variables X, Y, ... range over all finite subsets of this set, and $\in$ is always understood to denote the membership relation). The Archimedean postulate can be formulated and proves to be valid in $\mathscr{E}'_2$. Hence, by Theorem 1, every model of $\mathscr{E}'_2$ is isomorphic with a Cartesian space $\mathfrak{C}_2(\mathfrak{F})$ over some Archimedean real closed field $\mathfrak{F}$. There are, however, Archimedean real closed fields $\mathfrak{F}$ such that $\mathfrak{C}_2(\mathfrak{F})$ is not a model of $\mathscr{E}'_2$; e.g., the field of real algebraic numbers is of this kind. A consequence of the Archimedean postulate is that every model of $\mathscr{E}'_2$ has at most the power of the continuum (while, if only by virtue of Theorem 1, $\mathscr{E}_2$ has models with arbitrary infinite powers). In fact, $\mathscr{E}'_2$ has models which have exactly the power of the continuum, e.g., $\mathfrak{C}_2(\mathfrak{R})$, but it can also be shown to have denumerable models. Thus, although the theory $\mathscr{E}'_2$ may prove to be complete, it certainly has non-isomorphic models and therefore is not categorical.[8]

Only the decision problem for $\mathscr{E}'_2$ has found so far a definite solution:

THEOREM 5. *The theory $\mathscr{E}'_2$ is undecidable, and so are all its consistent extensions.*

This follows from the fact that Peano's arithmetic is (relatively) interpretable in $\mathscr{E}'_2$.[9]

To obtain the theory $\mathscr{E}''_2$ we leave the symbolism of $\mathscr{E}_2$ unchanged but we weaken the axiom system of $\mathscr{E}_2$. In fact, we replace the infinite collection of elementary continuity axioms, A13, by a single sentence, A13′, which is a consequence of one of these axioms. The sentence expresses the fact that

[8] These last remarks result from a general metamathematical theorem (an extension of the Skolem–Löwenheim theorem) which applies to all theories with the same logical structure as $\mathscr{E}'_2$, i.e., to all theories obtained from theories with standard formalization by including new variables ranging over arbitrary finite sets and a new logical constant, the membership symbol $\in$, and possibly by extending original axiom systems. By this general theorem, if $\mathscr{T}$ is a theory of the class just described with at most β different symbols, and if a mathematical system $\mathfrak{M}$ is a standard model of $\mathscr{T}$ with an infinite power α, then $\mathfrak{M}$ has subsystems with any infinite power γ, $\beta \leqslant \gamma \leqslant \alpha$, which are also standard models of $\mathscr{T}$. The proof of this theorem (recently found by the author) has not yet been published; it differs but slightly from the proof of the analogous theorem for the theories with standard formalization outlines in A. Tarski and R. L. Vaught, 'Arithmetical extensions of relational systems', *Compositio Mathematica*, Vol. 13 (1957), pp. 92 f. In opposition to theories with standard formalization, some of the theories $\mathscr{T}$ discussed in this footnote have models with an infinite power α and with any smaller, but with no larger, infinite power; an example is provided by the theory $\mathscr{E}'_2$ for which α is the power of the continuum. In particular, some of the theories $\mathscr{T}$ have exclusively denumerable models and in fact are categorical; this applies, e.g., to the theory obtained from Peano's arithmetic in exactly the same way in which $\mathscr{E}'_2$ has been obtained from $\mathscr{E}_2$. There are also theories $\mathscr{T}$ which have models with arbitrary infinite powers; such is, e.g., the theory $\mathscr{E}'''_2$ mentioned at the end of this paper.

[9] Cf. Tarski, Mostowski, and Robinson, *Undecidable Theories*, pp. 31 ff.

a segment which joins two points, one inside and one outside a given circle, always intersects the circle; symbolically:

$$\text{A13}'.\quad \bigwedge xyzx'z'u \bigvee y'[\delta(uxux') \wedge \delta(uzuz') \wedge \beta(uxz) \wedge \beta(xyz) \rightarrow \delta(uyuy') \wedge \beta(x'y'z')]$$

As a consequence of the weakening of the axiom system, various sentences which are formulated and valid in $\mathscr{E}_2$ are no longer valid in $\mathscr{E}_2''$. This applies in particular to existential theorems which cannot be established by means of so-called elementary geometrical constructions (using exclusively ruler and compass), e.g., to the theorem on the tri-section of an arbitrary angle.

With regard to metamathematical problems discussed in this paper the situation in the case of $\mathscr{E}_2''$ is just opposite to that encountered in the case of $\mathscr{E}_2'$. The three problems which are open for $\mathscr{E}_2'$ admit of simple solutions when referred to $\mathscr{E}_2''$. In particular, the solution of the representation problem is given in the following

THEOREM 6. *For $\mathfrak{M}$ to be a model of $\mathscr{E}_2''$ it is necessary and sufficient that $\mathfrak{M}$ be isomorphic with the Cartesian space $\mathfrak{C}_2(\mathfrak{F})$ over some Euclidean field $\mathfrak{F}$.*

This theorem is essentially known from the literature. The sufficiency of the condition can be checked directly; the necessity can be established with the help of the elementary geometrical theory of proportions (cf. the proof of Theorem 1).

Using Theorem 6 we easily show that the theory $\mathscr{E}_2''$ is incomplete, and from the description of $\mathscr{E}_2''$ we see at once that this theory is finitely axiomatizable.

On the other hand, the decision problem for $\mathscr{E}_2''$ remains open and presumably is difficult. In the light of the results of J. Robinson[10] it seems likely that the solution of this problem is negative; the author would risk the (much stronger) conjecture that no finitely axiomatizable subtheory of $\mathscr{E}_2$ is decidable. If we agree to refer to an elementary geometrical sentence (i.e., a sentence formulated in $\mathscr{E}_2$) as valid if it is valid in $\mathscr{E}_2$, and as elementarily provable if it is valid in $\mathscr{E}_2''$, then the situation can be described as follows: *we know a general mechanical method for deciding whether a given elementary geometrical sentence is valid, but we do not, and probably shall never know, any such method for deciding whether a sentence of this sort is elementarily provable.*

The differences between $\mathscr{E}_2$ and $\mathscr{E}_2''$ vanish when we restrict ourselves to universal sentences. In fact, we have

[10] 'Definability and decision problems in arithmetic', *Journal of Symbolic Logic*, Vol. 14 (1949), pp. 98–114.

THEOREM 7. *A universal sentence formulated in $\mathscr{E}_2$ is valid in $\mathscr{E}_2$ if and only if it is valid in $\mathscr{E}''_2$.*

To prove this we recall that every ordered field can be extended to a real closed field. Hence, by Theorems 1 and 6, every model of $\mathscr{E}''_2$ can be extended to a model of $\mathscr{E}_2$. Consequently, every universal sentence which is valid in $\mathscr{E}_2$ is also valid in $\mathscr{E}''_2$; the converse is obvious. (An even simpler proof of Theorem 7, and in fact a proof independent of Theorem 1, can be based upon the lemma by which every finite subsystem of an ordered field can be isomorphically embedded in the ordered field of real numbers.)

Theorem 7 remains valid if we remove A13′ from the axiom system of $\mathscr{E}''_2$ (and it applies even to some still weaker axiom systems). Thus we see that every elementary universal sentence which is valid in $\mathscr{E}_2$ can be proved without any help of the continuity axioms. The result extends to all the sentences which may not be universal when formulated in $\mathscr{E}_2$ but which, roughly speaking, become universal when expressed in the notation of Cartesian spaces $\mathfrak{C}_2(\mathfrak{F})$.

As an immediate consequence of Theorems 3 and 7 we obtain:

THEOREM 8. *The theory $\mathscr{E}''_2$ is decidable with respect to the set of its universal sentences.*

This means that there is a mechanical method for deciding in each particular case whether or not a given universal sentence formulated in the theory $\mathscr{E}''_2$ holds in every model of this theory.

We could discuss some further theories related to $\mathscr{E}_2$, $\mathscr{E}'_2$, and $\mathscr{E}''_2$; e.g., the theory $\mathscr{E}'''_2$ which has the same symbolism as $\mathscr{E}'_2$ and the same axiom system as $\mathscr{E}''_2$. The problem of deciding which of the various formal conceptions of elementary geometry is closer to the historical tradition and the colloquial usage of this notion seems to be rather hopeless and deprived of broader interest. The author feels that, among these various conceptions, the one embodied in $\mathscr{E}_2$ distinguishes itself by the simplicity and clarity of its metamathematical implications.

NOTES ON THE CONTRIBUTORS

Evert W. Beth (1908–64) was Professor of Logic, History of Logic, and Philosophy of Science at the University of Amsterdam. He wrote widely on all these subjects, as the bibliography of his works in *Synthese*, Vol. 16 (1966), pp. 90–106 shows. Professor Beth's best known work is probably *Foundations of Mathematics* (1959); his last book to be published in English was *Mathematical Thought* (1965).

Solomon Feferman is Professor of Mathematics and Philosophy at Stanford University. He is best known for his penetrating studies of the method of arithmetization and of the scope of predicative methods.

Kurt Gödel is Professor at the Institute of Advanced Study in Princeton. Although he has published, in addition to a few papers, only one work of monographic length (*The Consistency of the Continuum Hypothesis*, 1940), his thought has revolutionized nearly all branches of contemporary logic and foundational studies.

Leon Henkin is Professor of Mathematics at the University of California, Berkeley, and sometime (1962–4) President of the Association for Symbolic Logic. In addition to contributing to logic and foundational studies, Professor Henkin is co-author of *Retracing Elementary Mathematics* (1962)—'the first extensive treatment of the logical foundations of elementary mathematics ... accessible to readers without advanced graduate training'.

Georg Kreisel, F.R.S., is Professor of Logic and Mathematics in the Department of Philosophy of Stanford University, and is also Professor of Mathematics at the Université de Paris. In the last few years he has surveyed his own extensive work in logic and foundations, as well as the work of others, in several important survey articles and in his textbook (with J. L. Krivine) *Éléments de logique mathématique* (1967, English edition also 1967).

Abraham Robinson is Professor of Mathematics at Yale University. He is probably best known as one of the main architects of contemporary model theory (*On the Metamathematics of Algebra*, 1951; *Complete Theories*, 1956; *Introduction to Model Theory*, 1963) and as the main architect of non-standard analysis (*Non-Standard Analysis*, 1966).

Hartley Rogers, Jr., is Professor of Mathematics at the Massachusetts Institute of Technology. He has contributed to several branches of logic and foundational studies, especially to recursive function theory (*Theory of Recursive Functions and Effective Computability*, 1968). He is also known as an excellent expositor and teacher.

Raymond Smullyan teaches at Lehman College of the City University of New York. He is the author of *Theory of Formal Systems* (1961), *First-Order Logic* (1968), and numerous articles.

Alfred Tarski is Professor of Mathematics at the University of California, Berkeley. He is probably the most influential and prolific contemporary logician. (*Bibliography of the Writings of Alfred Tarski*, Department of Mathematics, University of California, Berkeley, 1965, lists 254 items.) Some of his earlier writings are collected in the volume *Logic*, *Semantics*, *Metamathematics* (1956). In addition to shaping large parts of modern logic and foundational studies, Tarski's work in semantics (model theory) has exerted a profound influence on philosophical discussions concerning the concept of truth and concerning the philosophy of mathematics.

BIBLIOGRAPHY

(*not including material in this volume*)

ANTHOLOGIES

Jean van Heijenoort, ed., *From Frege to Gödel. A Source Book in Mathematical Logic 1879–1931* (Harvard University Press, Cambridge, Mass., 1967).

Paul Benacerraf and Hilary Putnam, eds., *Philosophy of Mathematics: Selected Readings* (Prentice-Hall, Englewood Cliffs, N.J., 1964).

SURVEYS

Andrzej Mostowski, *Thirty Years of Foundational Studies: Lectures on the Development of Mathematical Logic and the Study of Foundations in 1930–1964*, Acta Philosophica Fennica, Vol. 17 (Basil Blackwell, Oxford, 1966).

Georg Kreisel, 'Mathematical Logic', in T. L. Saaty, ed., *Lectures on Modern Mathematics*, Vol. 3 (John Wiley & Sons, New York, London, and Sydney, 1965, pp. 95–195).

GENERAL WORKS ON MATHEMATICAL LOGIC AND THE FOUNDATIONS OF MATHEMATICS

David Hilbert and Paul Bernays, *Grundlagen der Mathematik I–II* (Springer-Verlag, Berlin, 1934–9).

S. C. Kleene, *Introduction to Metamathematics* (Van Nostrand, New York, 1952).

Alfred Tarski, *Logic, Semantics, Metamathematics: Papers from 1923 to 1938*, translated by J. H. Woodger (Clarendon Press, Oxford, 1956).

Evert W. Beth, *The Foundations of Mathematics* (North-Holland Publishing Company, Amsterdam, 1959; second edition, Harper & Row, New York, 1966).

Joseph R. Shoenfield, *Mathematical Logic* (Addison-Wesley Publishing Company, Reading, Mass., 1967).

PHILOSOPHICAL DISCUSSIONS OF LOGIC AND THE FOUNDATIONS OF MATHEMATICS

Kurt Gödel, 'Russell's Mathematical Logic', in Paul A. Schilpp, ed., *The Philosophy of Bertrand Russell* (The Library of Living Philosophers,

Evanston, Ill. (later, Open Court Publishing Company, La Salle, Ill.), 1944), pp. 125–53.

Leon Henkin, 'Some Notes on Nominalism', *Journal of Symbolic Logic*, Vol. 18 (1953), pp. 19–29.

Ludwig Wittgenstein, *Remarks on the Foundations of Mathematics*, ed. G. H. von Wright and G. E. M. Anscombe and translated by G. E. M. Anscombe (Basil Blackwell, Oxford, 1961).

Abraham Robinson, 'Formalism 1964', in Yehoshua Bar-Hillel, ed., *Logic Methodology and Philosophy of Science*, Proceedings of the 1964 International Congress (North-Holland Publishing Company, Amsterdam, 1965), pp. 228–46.

W. V. O. Quine, *The Ways of Paradox and Other Essays* (Random House, New York, 1966).

Hilary Putnam, 'Mathematics without Foundations', *Journal of Philosophy*, Vol. 64 (1967), pp. 5–22.

Natural Deduction Methods and Other Developments in First-Order Logic

J. Herbrand, *Écrit logiques*. J. van Heijenoort, ed. (Presses Universitaires de France, Paris, 1968).

G. Gentzen, 'Untersuchungen über das logische Schliessen I–II', *Mathematische Zeitschrift*, Vol. 39 (1934), pp. 176–221, 403–31. Reprinted in French in: G. Gentzen, *Recherches sur la déduction logique*, R. Feys and J. Ladrière, eds. (Paris, 1955) and in English in *American Philosophical Quarterly*, Vol. 1 (1964), pp. 288–306, and Vol. 2 (1965), pp. 204–18.

H. Rasiowa and R. Sikorski, 'On the Gentzen Theorem', *Fundamenta Mathematicae*, Vol. 58 (1960), pp. 59–69.

Jaakko Hintikka, 'Form and Content in Quantification Theory', *Acta Philosophica Fennica*, Vol. 8 (1955), pp. 11–55.

Kurt Schütte, 'Ein System des verknüpfenden Schliessens', *Archiv für mathematische Logik und Grundlagenforschung*, Vol. 2 (1956), pp. 55–67.

William Craig, 'Linear Reasoning: A New Form of the Herbrand-Gentzen Theorem', *Journal of Symbolic Logic*, Vol. 22 (1957), pp. 250–68.

Jaakko Hintikka, 'Distributive Normal Forms in First-Order Logic', in J. N. Crossley and M. A. E. Dummett, eds., *Formal Systems and Recursive Functions*, Proceedings of the Eighth Logic Colloquium, Oxford, July

1963 (North-Holland Publishing Company, Amsterdam, 1965), pp. 47-90.

Model Theory

Alfred Tarski, 'Contributions to the Theory of Models I–III', *Indagationes Mathematicae*, Vol. 16 (1954), pp. 572–81, 582–8, and Vol. 17 (1955), pp. 56–84.

J. W. Addison, Leon Henkin, and Alfred Tarski, eds., *The Theory of Models*, Proceedings of the 1963 International Symposium at Berkeley (North-Holland Publishing Company, Amsterdam, 1965).

Abraham Robinson, *Introduction to Model Theory and to the Metamathematics of Algebra* (North-Holland Publishing Company, Amsterdam, 1963).

William Craig, 'Three Uses of the Herbrand-Gentzen Theorem in Relating Model Theory and Proof Theory', *Journal of Symbolic Logic*, Vol. 22 (1957), pp. 269–85.

Robert L. Vaught, 'Models of Complete Theories', *Bulletin of the American Mathematical Society*, Vol. 69 (1963), pp. 299–313.

Robert L. Vaught, 'The Löwenheim-Skolem Theorem', in Yehoshua Bar-Hillel, ed., *Logic, Methodology and Philosophy of Science*, Proceedings of the 1964 International Congress (North-Holland Publishing Company, Amsterdam, 1965, pp. 81–9).

Proof Theory

Kurt Schütte, *Beweistheorie* (Springer-Verlag, Berlin, Göttingen, Heidelberg, 1960).

Dag Prawitz, *Natural Deduction: A Proof-Theoretical Study*, Stockholm Studies in Philosophy, Vol. 3 (Almqvist & Wiksell, Stockholm, 1965).

Georg Kreisel, 'A Survey of Proof Theory', *Journal of Symbolic Logic*, Vol. 33 (1968), pp. 321–388.

Incompleteness and Undecidability

See also the works listed under 'Recursive Function Theory', especially Rogers and the papers reprinted in Davis.

A. Tarski, A. Mostowski, and R. M. Robinson, *Undecidable Theories* (North-Holland Publishing Company, Amsterdam, 1953).

Hao Wang, 'Undecidable Sentences Created by Semantic Paradoxes', *Journal of Symbolic Logic*, Vol. 20 (1955), pp. 31–43.

Solomon Feferman, 'Arithmetization of Metamathematics in a General Setting', *Fundamenta Mathematicae*, Vol. 49 (1960–1), pp. 35–92.

Higher-Order Logics

Kurt Schütte, 'Syntactical and Semantical Properties of Simple Type Theory', *Journal of Symbolic Logic*, Vol. 25 (1960), pp. 305–26.

Richard Montague, 'Set Theory and Higher-Order Logic', in J. N. Črossley and M. A. E. Dummett, eds., *Formal Systems and Recursive Functions*, Proceedings of the Eighth Logic Colloquium, Oxford, July 1963 (North-Holland Publishing Company, Amsterdam, 1965), pp. 131–48.

Dag Prawitz, 'Hauptsatz for Higher Order Logic', *Journal for Symbolic Logic*, Vol. 33 (1968), pp. 452–57.

Predicativity

Solomon Feferman, 'Autonomous Transfinite Progressions and the Extent of Predicative Mathematics', in B. van Rootselaar and J. F. Staal, eds., *Logic, Methodology and Philosophy of Science III*, Proceedings of the 1967 International Congress (North-Holland Publishing Company, Amsterdam, 1968, pp. 121–35).

Georg Kreisel, 'La prédicativité', *Bulletin de la Société Mathématique de France*, Vol. 88 (1960), pp. 371–91.

Set Theory

Kurt Gödel, *The Consistency of the Axiom of Choice and of the Generalized Continuum-Hypothesis with the Axioms of Set Theory*, Annals of Mathematics Studies Vol. 3 (Princeton University Press, Princeton, N.J., 1940).

Kurt Gödel, 'What Is Cantor's Continuum Hypothesis?', *American Mathematical Monthly*, Vol. 54 (1947), pp. 515–25. Reprinted with additions in Benacerraf and Putnam, eds., *Philosophy of Mathematics, Selected Readings* (Prentice-Hall, Englewood Cliffs, N.J., 1964), pp. 258–73.

Paul Cohen, *Set Theory and the Continuum Hypothesis* (W. A. Benjamin, New York and Amsterdam, 1966).

W. V. O. Quine, *Set Theory and Its Logic* (Harvard University Press Cambridge, Mass., 1963).

Andrzej Mostowski, 'Recent Results in Set Theory', in I. Lakatos, ed. *Problems in the Philosophy of Mathematics*, Proceedings of the International Colloquium in the Philosophy of Science (London, 1965), Vol. 1 pp. 82–96 (discussion, pp. 97–118).

A. A. Fraenkel and Y. Bar-Hillel, *Foundations of Set Theory* (North Holland Publishing Company, Amsterdam, 1958).

Dana Scott, 'A Proof of the Independence of Continuum Hypothesis' *Mathematical Systems Theory*, Vol. 1 (1966), pp. 89–111.

RECURSIVE FUNCTIONS AND COMPUTABILITY

Martin Davis, ed., *The Undecidable: Basic Papers on Undecidable Propositions, Unsolvable Problems, and Computable Functions* (Raven Press Hewlett, New York, 1965).

Hartley Rogers, Jr., *Theory of Recursive Functions and Effective Computability* (McGraw-Hill, New York, 1968).

Martin Davis, *Computability and Unsolvability* (McGraw-Hill, New York 1958).

M. O. Rabin and D. Scott, 'Finite Automata and Their Decision Problems' *IBM Journal of Research and Development*, Vol. 3 (1959), pp. 114–25.

INTUITIONISM

L. E. J. Brouwer, 'Historical Background, Principles and Methods of Intuitionism', *South African Journal of Science*, Vol. 49 (1952), pp. 139–46

A. Heyting, *Intuitionism: An Introduction* (North-Holland, Amsterdam 1956).

A. Heyting, 'After Thirty Years', in Ernest Nagel, Patrick Suppes, and Alfred Tarski, ed., *Logic, Methodology and Philosophy of Science* Proceedings of the 1960 International Congress (Stanford University Press, Stanford, Calif., 1962).

Evert W. Beth, 'Semantic Construction of Intuitionistic Logic', *Mededelingen van de Koninklijke Nederlandse Akademie van Wetenschappen*, Afd Letterkunde, N.R. Vol. 19, no. 11 (1956), pp. 357–88.

V. H. Dyson and Georg Kreisel, *Analysis of Beth's Semantic Construction of Intuitionistic Logic*. Applied Mathematics and Statistics Laboratory Technical Report no. 3 (Stanford University, Stanford, Calif., 1961).

ul Kripke, 'Semantical Analysis of Intuitionistic Logic I', in J. N. Crossley and M. A. E. Dummett, ed., *Formal Systems and Recursive Functions*, Proceedings of the Eighth Logic Colloquium, Oxford, July 1963 (North-Holland, Amsterdam, 1965, pp. 92–130).

C. Kleene and R. E. Vesley, *The Foundations of Intuitionistic Mathematics* (North-Holland, Amsterdam, 1965).

ONSTRUCTIVE AND FINITISTIC FOUNDATIONS (OTHER THAN INTUITIONISM)

urt Gödel, 'Über eine bisher noch nicht benützte Erweiterung des finiten Standpunktes', in *Logica: Studia Paul Bernays Dedicata* (Editions Griffon, Neuchatel, 1959), pp. 76–83.

ul Lorenzen, 'Ein dialogisches Konstruktivitätskriterium', in *Infinitistic Methods*, Proceedings of the Symposium on Foundations of Mathematics (Pergamon Press, London, and Państwowe Wydawnictwo Naukowe, Warszawa, 1961), pp. 193–200.

ON-STANDARD ANALYSIS

braham Robinson, *Non-Standard Analysis* (North-Holland, Amsterdam, 1966).

ATHEMATICAL DISCOVERY. THE DEVELOPMENT OF MATHEMATICS

. Polya, *Mathematical Discovery* I–II (John Wiley, New York, 1962–5).

re Lakatos, 'Proofs and Refutations', *The British Journal for the Philosophy of Science*, Vol. 14 (1963), pp. 1–25, 120–39, 221–45, 296–342.

EOMETRY

avid Hilbert, *Grundlagen der Geometrie*, 8th ed., with revisions and supplements by P. Bernays, Stuttgart, 1956.

olfram Schwabhäuser, 'Metamathematical Methods in Foundations of Geometry', in Yehoshua Bar-Hillel, ed., *Logic, Methodology and Philosophy of Science*, Proceedings of the 1964 International Congress (North-Holland, Amsterdam, 1965), pp. 152–65.

Mathematics and Logic

Leon Henkin, 'Are Logic and Mathematics Identical?', *Science*, Vol. 13 no. 3542 (16 November 1962), pp. 788–94.

Hilary Putnam, 'The Thesis that Mathematics is Logic', in R. Schoenma ed., *Bertrand Russell: Philosopher of the Century* (George Allen & Unwi London and Atlantic-Little, Brown & Co., Boston, 1967), pp. 273–303

The Status of Mathematical and Logical Truths

W. V. O. Quine, 'Carnap and Logical Truth', in P. A. Schlipp, ed., *T Philosophy of Rudolph Carnap* (Library of Living Philosophers, Vol. 1 Open Court, La Salle, Ill., 1963), pp. 385–406.

Jaakko Hintikka, 'Are Logical Truths Analytic?' *Philosophical Revie* Vol. 74 (1965), pp. 178–203.

INDEX OF NAMES

(not including authors mentioned only in the Bibliography)